ASHEVILLE-BUNCOMBE TECHNICAL INSTITUTE NORTH CAROLINA
694 NE STATE BOARD OF EDUCATION
DEPT. OF COMMUNITY COLLEGES
LIBRARIES

11-73

DISCARDED

JUN 1 2 2025

Henry Laurence Gantt

Leader in Industry

Henry Laurence Gantt

LEADER IN INDUSTRY

BY

L. P. ALFORD

EASTON

HIVE PUBLISHING COMPANY

1972

This is a facsimile reprint of the 1934
edition published in New York by Harper & Brothers.

Hive Management History Series: No. 6

Library of Congress Catalog Card Number 79-169661

Standard Book Number 0-87960-000-4

Contents

Preface	ix
PART I: PREPARATION	
Foreword	3
I. The Plantation in Maryland	5
II. McDonogh Days	17
III. Johns Hopkins and Stevens	35
IV. Homes and Family	46
PART II: ACTION	
Foreword	59
V. Technical Engineering	61
VI. Scientific Management	78
VII. Task and Bonus	85
VIII. Sayles Bleacheries	107
IX. Sayles Bleacheries (Continued)	126
X. Training Workmen	138
XI. Remington Typewriter	149
XII. Remington Typewriter (Continued)	158
XIII. Cheney Brothers	166
XIV. The Years of War	185
XV. The Gantt Chart	207
XVI. Pine Island Farm	224
XVII. End of a Life of Service	236

CONTENTS

PART III: VISION

Foreword	247
XVIII. Democracy in Industry	251
XIX. The New Machine	264
XX. The Parting of the Ways	278
XXI. The Motive of Service	288
Appendix	300
Index	311

Illustrations

H. L. Gantt	*Frontispiece*	
Virgil and Mary Gantt	*Facing page*	10
Henry and Margaret Gantt	" "	10
The Plantation Where Gantt Was Born	" "	14
The "Little Red Schoolhouse" Where Gantt's First Days of Learning Were Spent	" "	14
"Oakleigh"	" "	14
Mary Eliza Snow at the Time of Her Engagement	" "	50
Gantt, Peggy and Roy	" "	50
A "Red-and-Black" Bonus Chart for a Weave-room	*Page*	98
A Daily Graphical Balance Chart	"	103
Letter of Appreciation from the Foremen of the Schenectady Works of the American Locomotive Company	"	105
A Percentage Chart	"	119
Gantt Chute or Automatic Piling Machine	*Facing page*	120
Cost and Production Charts Developed at Remington	*Page*	164
Idleness Expense Chart	"	183
First Gantt Chart Plotted for Artillery Ammunition	"	210
First Gantt Chart to be Published	"	212

ILLUSTRATIONS

TITLE PAGES OF VARIOUS EDITIONS OF "THE GANTT CHART"	*Page* 216
THE HENRY LAURENCE GANTT MEMORIAL GOLD MEDAL	*Facing page* 240
CERTIFICATE ACCOMPANYING THE MEDAL AWARDED POSTHUMOUSLY TO GANTT	” ” 242

PREFACE

THIS biography of Henry L. Gantt has been written at the initiative and under the auspices of the Biography Committee of the American Society of Mechanical Engineers. The specifications of that body as to what an engineer's biography "should contain" have guided the treatment adopted for the work. Following out these requirements there are flashes of the formative forces, convictions, and influences that shaped Gantt, details of his engineering achievements, and a summary of his philosophic teachings.

The narrative of an engineer's life, as objectified by the Biography Committee, "should be so planned that the story of his inheritance, education, and growth to manhood and citizenship should show the manner of man" he was. ". . . his tastes and attainments, how he fitted into his place in this workaday world, and the ambitions and aspirations which led him to attempt the profession and art of engineering" should be presented and explained.

As to his professional life: "The account of his practice should detail the main lines of his activities with their failures or successes, giving details where possible, and the additions and modifications which he brought to the art and practice of his profession. His growth as a citizen should be thrown into juxtaposition with his professional life and his influence on his older and younger confrères should be emphasized.

"Finally, his influence on the trend of engineering should

be evaluated, followed by an estimation of the extent and scope of his influence on the separate lines in which he worked and the bearing of these facts on the future of engineers and the engineering profession. The biography should conclude with an appendix containing a good chronology of employment, work done, papers and books, offices, membership, and honors."

The exploration for facts and information to satisfy this circumstantial outline was made among members of his family, classmates, clients, professional associates, friends, and admirers. Fortunately, a large number of articles and papers written by Gantt were discovered. The most serious lack in this respect was his correspondence and reports; his office files could not be traced. In addition to these documents considerable anecdotal and reminiscent material was secured. Library research brought to light other papers and articles published in technical and popular magazines. His books, three in number, were invaluable sources of material.

An effort has been made to check all the data and information in this biography, by having every chapter read by one or more individuals who had some personal knowledge of the events described. It is futile, even with care of this kind, to expect that a work dealing with such an active life can be entirely accurate or wholly complete. There will be blunders and omissions no matter how much pains have been taken to avoid them.

The arrangement adopted divides the book into three parts which are distinct in subject-matter, although overlapping in time. The divisions are named: Preparation; Action; Vision. In brief, Part I is devoted to Gantt's heredity, education, and formative environments. Part II deals with

PREFACE

the activities and accomplishment, successes and failures, of his professional life. Part III presents his industrial philosophy.

In a real sense this biography is the work of relatives, friends, associates, and admirers. They have supplied the information and much of the interpretation. It is a pleasant duty to acknowledge their generous help and to express appreciation for the cooperative spirit in which this assistance has been offered. To name those who assisted in this way:

William S. Aldrich
Grace Hutchins Alford
Charles Whiting Baker
Carl L. Bausch
Carle M. Bigelow
Ellen Snow Brown
Reverend Francis A. Brown
Janet Allan Bryan
Roy I. Bull
H. M. Burke
Horace B. Cheney
Pearl F. Clark
Wallace Clark
Colonel W. L. Conrad
General William Crozier
J. L. Cox
C. E. Davies
George A. Dornin
Alfred F. Ernst
H. R. Fardwell
D. C. Fenner

George M. Forrest
Col. Benjamin A. Franklin
John P. Frey
R. S. Gardner
Thayer P. Gates
S. R. Gerber
Dr. Lillian M. Gilbreth
Charles B. Going
Reverend Carl E. Grammer
H. Steuart Jackson
Mary Steuart Jackson
W. O. Jelleme
Dean Dexter S. Kimball
Major Louis E. Lamborn
Conrad N. Lauer
Gordon Lee
Samuel H. Libby
Reynolds Longfield
William L. Lyall
Professor D. C. Lyle
Professor C. W. Lytle

F. D. Manning
E. P. Miller
Major Fred J. Miller
Theodore H. Miller
Harvey F. Mitchell
Colonel George Montgomery
J. W. Nickerson
Edward B. Passano
Colonel J. H. Pelot
Thomas Pettigrew
Walter N. Polakov
Professor David B. Porter
Dr. Harry Fielding Reid
H. P. Reno
Calvin W. Rice

Professor Joseph W. Roe
Dean Herman Schneider
Ralph Shaw
Henry H. Suplee
George Z. Sutton
Margaret Gantt Taber
Thomas T. Taber
Captain W. C. Thee
C. W. Thomas
General John T. Thompson
Charles N. Underwood
R. A. Wentworth
Roy V. Wright
D. Robert Yarnall

Gratefulness is also due to Agnes M. McTernan for her editorial assistance, which included all the work of library research.

The foregoing acknowledgment of indebtedness to others is wholly inadequate, for this biography could not have been written without their assistance. It must be recognized, however, that for the exploratory search for information, the selection of material to be included, the general manner of presentation, and for all statements made, I am solely responsible.

In discharging this responsibility I have been interested only in the truth so far as it could be determined and written down, in some instance more than a half-century after the event. I have tried:

PREFACE

To be quite fair, honestly critical, and completely sympathetic.

To present Gantt as a human personage with weaknesses and shortcomings, but with a power and capacity for leadership which were felt by all who met him.

To make Gantt come alive.

L. P. Alford

Montclair, New Jersey
August, 1934

Part I

PREPARATION

FOREWORD

Prominent among the directive forces and influences that mold character and prepare for a life of work are a man's home and parents, his schools and teachers, and his implanted ideals and honored heroes. Events of the times, existing economic conditions, and possible opportunities may also play a recognizable part in a boy's development to the strength of mature manhood. These formative agencies are included under the term environment. Back of them in relation to time, are the inbred traits and abilities which have their origin in the family tree.

In the record of the life of Henry L. Gantt forces and capabilities appear which came to him through long lines of able ancestors. These heritages, elements of character and capacity, were his just due. There can also be determined the effect upon his surroundings and conditions of life of the social and political events of the adventurous years of his boyhood. The rigorous, strenuous period spanned by his manhood included equally, eventful times, but with a different resultant. Building from this information, Part I of this biography is devoted to his heredity, the environment of boyhood and youth, and the homes of his later years.

It is, in its opening pages, a dramatic story from American life of the second half of the nineteenth century. The narrative passes from the shadows of hardship and struggle to the sunshine of winnings and success. The Civil War changed Gantt's environments. Family disaster swept away hereditary privileges and brought meagerness of surroundings. Education was gained through his own efforts. High scholastic honors were won. The preparation and training thus secured in spite of many difficulties were better than the average enjoyed by the engineers of his day. In pleasant contrast with this period of struggle his life in his own homes during the first quarter of the twentieth century was placid and serene.

Another feature in this story is the influence of outstanding personalities upon the developing boy and youth. Some were national figures, men of position, importance, and accomplishment. They brought constructive direction to bear upon the methods and quality of his thinking, and aided in shaping his ideals, principles, and convictions.

The four chapters making up Part I, taken together, present those features and events in Gantt's life, other than certain factors of a professional nature, which tended to make him what he was, a Leader in Industry.

Chapter I

THE PLANTATION IN MARYLAND

On April 12, 1861, Confederate guns roared out in bombardment of Fort Sumter. After thirty hours of attack the fortress was compelled to surrender. The next day President Lincoln's arousing call for volunteers flashed over a hitherto happy country pictured by Lowell as in easeful repose: "America lay asleep, like the princess of the fairy tale, enchanted by prosperity. But at the fiery kiss of war the spell is broken, the blood tingles along her veins, and she awakens, conscious of her beauty and her sovereignty." Thus came the war between the North and the South. The one fought to "preserve the Union"; the other struggled to "keep the past upon its throne."

Thirty-six days after Sumter surrendered, a son was born to Virgil and Mary Gantt. They named him Henry Laurence.

The place of his birth was a plantation of one of the "border states," Maryland, with wide-flung fertile fields sweeping down to a broad, lazy, tidal river. Across those acres for generations the work songs of negro slaves had been heard as they planted, tended, and harvested the abundant crops of cotton, corn, and tobacco. At the time of his birth his family was enjoying the peace, plenty, and prosperity of the country. A happy household group of parents and sister welcomed the newcomer; the home possessed an abundance of the good things of life; the family was affluent, if not wealthy. All this fortune was soon to change. The war just begun was destined to have a far-reaching effect upon the boyhood and youth of the sturdy baby who had just been born. A series of disastrous events reduced the family of Virgil Gantt to near poverty within a short decade after Laurence's birth.

From twelve years of age on Gantt made his own way in life through sheer ability, pitting himself at first against conditions caused by one of the worst of American business panics. He put himself through a university and a technical institute to become one of the most eminent engineers of his times, a patient teacher of better ways of doing work, a courageous leader in industrial improvement, and a philosopher of a new business and social structure. Along the way in youth he won prizes, honors, and scholarships; during his manhood years he earned distinction through the formulation of principles, for the invention of mechanisms, and by substantial, enduring achievements in the newest branch of engineering. These contributions will remain so long as men are organized to do productive work. The ever-flowing stream of human experience, enriched, deepened, and widened by his life and accomplishments, passed on from him a heightened heritage for all who follow.

Those who hold to the truth of the maxim, "blood will tell," will find satisfaction in Henry Laurence Gantt's lineage. In the ancestral line whence he came, beginning with Thomas Gantt who emigrated from England, are a clergyman of the Episcopal Church, a physician, two men of affairs, and a soldier. Tracing farther back into the records of the family before the migration to the Colonies, one finds

THE PLANTATION IN MARYLAND 7

another clergyman, a rector of Dudley, England, and another soldier. The Gantt family claims to be descended from Gilbert de Gantt, son of Baldwin, Earl of Flanders, who joined the expeditionary force of William the Conqueror to England in 1066, fought in the battle of Hastings, and established the Gantt family on British soil.

To sketch the family line with more detail, Thomas Gantt and his Scotswoman wife, Mary Graham, emigrated to the colony of Maryland in 1660 and settled on a plantation named "Myrtle Range" in Prince George's County. The location was near White Landing on the Patuxent River. Here the family lived the life of pioneer planters, Thomas being active in public affairs. In 1683 he was a Justice of the Quorum. In 1689 he was one of His Majesty's Justices, and one of a committee to regulate the civil affairs in Calvert County. The lands and locality he selected as the home spot where he would establish the Gantt family in Colonial America were pleasant indeed. The open meadows, tree-crested knolls, and swampy stretches beside the streams, became landmarks for his children, and children's children, for more than two centuries of independent and happy country life.

Thomas Gantt's son, Edward, made his home a short distance to the southward from the place where the family had first settled, in adjacent Calvert County, and to the eastward of the Patuxent River. Here a son was born named Thomas after his grandfather. In 1715 "Myrtle Range," his grandfather's plantation, was reserved for him. He became a man of importance in the growing colony. In 1725 he was a Justice, in 1728 a Justice of the Quorum, and in 1732 a Justice of the Peace. In discharging the duties of those public offices he was following in the footsteps, emulating the example, and serving his day and generation, much as his emigrant ancestor had done before him. The willingness for public service which he thus showed exhibited itself again and again in the male line leading down to Henry Laurence Gantt.

Continuing to trace the lineage, the next is Captain Edward Gantt, the great-great-grandfather of Henry Laurence. He was a soldier and a churchman. In 1752 he was Captain of the Colonial Troops, and in 1776 Captain in the Continental Army. In addition, he was a member of the Association of Freemen, a vestryman of Christ Church, Calvert County in 1747, and later of All Saints' Church in the same county. Through him the family continued its contact with public affairs and participated in the War for Independence. The spirit of this liberty-loving patriot was to reappear again, at the time of another national emergency in America, in his great-great-grandson who gave himself unsparingly in the service of his country.

Two professional men now appear, father and son. The Reverend Edward Gantt, clergyman of the Church of England, was Henry Laurence's great-grandfather, and his son was Dr. Thomas Compton Gantt, a physician and surgeon. Reverend Gantt, in 1784, was ordained deacon by the Bishop of Chester, and priest by the Archbishop of Canterbury. He was one of the first clergymen from Maryland to be ordained in England. His warrants of ordination were issued under an Act of the British Parliament whereby the candidate for ordination was not required to take an oath of allegiance to the British Crown, but was restricted in conducting his clerical offices to countries outside of the British

THE PLANTATION IN MARYLAND 9

Dominions. These two members of learned professions show across the years a high level of intellectual powers, on the one hand, and, on another, the fortunate economic situation of their families which permitted higher education and foreign travel.

The ancestral line now comes to the immediate family of Henry Laurence Gantt. Virgil Gantt, his father, a son of Dr. Thomas Compton Gantt, was born in 1820. He was distinctly the southern planter of his times. He was educated at Charlotte Hall in St. Mary's County, Maryland, and at St. John's College in Annapolis. His life, until the coming of economic misfortune, was spent on the plantation where he was born. His activities were the managing of the estate and, during the season when work was active on the farm lands, much of every day was spent by him in the saddle riding from one group of slaves to another, supervising and overseeing all that was being done. From early manhood he suffered a physical disability, deafness. This was a major handicap and resulted in throwing an increased burden of responsibility upon his beloved wife, Mary.

In his home life he was genial and pleasant, and took great delight in playing with his children. He would "ride" them on his knee or foot, and often would impersonate "the big bear" by dropping to his hands and knees and growling, to the extravagant ecstasy of the little folks. When the children were old enough he would take them on horseback with him, frequently placing little Mary in front of the saddle and Laurence behind.

Mary Jane Steuart, whom Virgil Gantt married November 14, 1855, was a Baltimore city girl. She was born in 1832 and thus was twelve years his junior. She was educated at the Snowden School for Girls. When only sixteen, due to her mother's ill health, she took over the management of her father's house, being the eldest daughter. In her mature years she was a fine-looking woman, with such charm of manner and vividness of personality that whenever she entered a room everyone instinctively turned to her. Her character was strong, honest, and upright. She spoke what she believed without fear or favor, and perhaps somewhat abruptly at times.

She was devoted to her church. Every morning she read her Bible, and every Sunday afternoon for years she gathered her grandchildren together in a class to study the Catechism and Creed, and, after the lesson, to have tea with Grandma. A woman of the old South in outlook, belief, and feeling, she found it difficult to adjust herself to the changes in family fortune, but did her part with heroic fortitude. From her Laurence inherited much.

Though an invalid for many years, Mary Steuart Gantt was never idle. Her industry and charity are both shown by her activity during the World War. During 1917-1918 she knitted with her own hands sixty-nine sweaters, sixty-five pairs of socks, and numerous other articles for our soldiers in France.

Eight children were born to Virgil and Mary Steuart Gantt, of whom four died in infancy. Two girls and two boys lived to womanhood and manhood. Of these the eldest, and the first-born child, was Margaret Heighe, born in 1856, deceased in 1888. The eldest of the surviving boys, the fourth child, was Henry Laurence, born on May 20, 1861. Then followed Mary Steuart, the well-beloved sister of

Henry and Margaret Gantt

Virgil and Mary Gantt

THE PLANTATION IN MARYLAND

Gantt's mature years. The youngest of this group of four children was Charles Steuart.

The plantation near Prince Fredericktown, the county seat of Calvert County, Maryland, where Henry Laurence was born, extended for more than a mile along the shores of the Patuxent River. It was a pleasant stretch of country and the plantation produced large crops of staples, principally tobacco. These were shipped to commission houses in Baltimore for sale, and from that city were procured the numerous supplies necessary to keep a large plantation operating. The plantation home was a manor house built of English brick with a floor plan in the shape of a cross, a by no means unusual design for pre-Civil War, Southern dwellings. Adjacent to the master's house were three other brick buildings—the coachman's house, the dairy, and the smoke-house. At the foot of the broad sweeping lawn near the road was the office of Dr. Thomas C. Gantt, a delightful place for Laurence and Mary to play. Near the road also were the wide-flung barns, and behind the house on the top of a little hill were the negro cabins. Some three miles away was St. Paul's Church, where many of the church-going, God-worshiping ancestors of Henry Laurence had been communicants. In the churchyard at the side a number of the members of his immediate family were buried; others were at rest in a private cemetery on the plantation. A short distance away down the country road was a small wooden schoolhouse where his first days of learning were spent. Each of these institutions, the church, and the school, made its impression upon the boy Laurence.

Frequently on Sunday afternoon, after he had attended church in the morning, he would gather together a congregation composed of his sister Mary and his colored nurse, Mary Ellen. He would then slip his long white nightshirt over his Sunday clothes, stand on a chair, and "preach" to his audience of two.

An incident from the school is of quite a different character. The teacher was a man who had the unfortunate habit of frequently drinking too much liquor. On one such occasion he accused Laurence of lying. The hot-tempered, honest youngster flew into a rage, seized a bootjack that was lying on the floor, and threw it at the schoolmaster's head. That act resulted in his withdrawal from school. His mother took both him and his sister Mary out of that school and thereafter, so long as they lived on the plantation, taught them herself at home.

In earlier paragraphs a hint has been given of a series of events which overtook the Virgil Gantt family as a result of the war between the North and the South. Maryland was one of the "border states." Because of the skilled diplomacy of President Lincoln and the coercive force of Union armies, Maryland was held to the cause of the North. However, many of her young men joined regiments in the Confederate army, and many of her people were totally and unalterably sympathetic with the Confederacy. Among the latter were Virgil and Mary Gantt.

From almost the outbreak of hostilities a Union army was encamped at Benedict, Maryland, only about a mile from the Gantt plantation house across the Patuxent River. For a while, and in spite of all personal leanings and external events, the family life moved along without serious disturbance. Finally the first blow struck. On September 23, 1862, President Lincoln issued the Emancipation Proclama-

THE PLANTATION IN MARYLAND 13

tion which was to take effect on the first day of the following January. In this he specifically exempted from its provisions the border states and such parts of the Southern states as were under the control of the Union armies. For this reason the proclamation did not emancipate the slaves in the State of Maryland. However, so far as Virgil Gantt was concerned the effect upon his fortunes was immediate. Under the influence and persuasion of soldiers in the Union army encamped at Benedict his sixty-nine slaves deserted. In fact, the Union soldiers practically kidnapped the male slaves at night, taking them to their camp as servants and laborers. A few came back, but the going of the majority represented a major property loss, probably some forty to fifty thousand dollars. Somewhat later, Maryland, by a special Act of the Legislature, emancipated all of the slaves within the state without compensation to their owners.

Then came months of difficulty under new economic conditions. Slave labor on a subsistence level of living was replaced by free labor demanding wages and aspiring to a higher and increasing standard of living. The change to a Southern planter was a major overturn in both economic outlook and operating method, far more drastic than has been experienced by any one due to a business panic or an industrial depression. For Virgil Gantt the difficulties increased; one unfortunate event after another happened until in 1871 the mortgage on his plantation was foreclosed and the family was compelled to move away from the locality where their people had prospered for more than two hundred years. In the city of Baltimore they established a new home, far different from the one they had loved and enjoyed on the ancestral acres.

At this time the family consisted of seven members— Virgil and Mary Gantt, the parents, Grandmother Margaret Heighe Steuart, and the children, Margaret, Henry Laurence, Mary, and Charles. The physical disability of deafness prevented Virgil Gantt from doing much to help his family from this time on. In this situation the wife and mother took upon herself the major responsibility, and by running a boarding-house kept the family together. Margaret helped somewhat later when she had become a schoolteacher. Southern in feeling as Mary Steuart Gantt was, and resenting as she did the war and its outcome, she suffered as did many a mother of the old South, feeling keenly the turn of events which had reduced her from a position of affluence and social prestige to one where she had to deal single-handed with the every-day, stubborn problems of existence.

The strain was more than her health could stand for long. Early in 1883 Gantt wrote to Dr. D. C. Gilman, asking advice about his own future, and added, by way of explaining his circumstances, "My mother, who to within a year had supported herself by keeping a few boarders, has now become an invalid."

The economic results of the Civil War as it affected the fortunes of his family, and the circumstances of his boyhood and youth, left impressions that continued with him throughout his entire life. In some degree his sympathy for the unfortunate, such as the Belgian refugees of the World War, can be attributed to these experiences. His personal frugality and dislike of ostentatious spending undoubtedly had their rise in his own privations. His realization that economic disaster can overtake an entire people drew its

The Plantation Where Gantt Was Born

The "Little Red Schoolhouse" Where Gantt's First Days of Learning Were Spent

"Oakleigh"

strength and vividness not only from intimate observation, but from personal trial.

A little incident that occurred in Baltimore shows the restless eagerness of the boy to learn by conscious experience, a bent of his mind which later on became one of his outstanding characteristics, and gave rise to his continued questioning of all things that came before him. At the time of this happening the family was attending St. Luke's Church. Laurence said he wanted to see how it felt to be a choir boy. So one Sunday he was permitted, to his deep satisfaction, to put on the vestments, cassock and cotta, march in with the choir, and do as the other boys did throughout the service—except he did not sing. He could not sing, but he was a choir boy for a morning.

Virgil Gantt died nine years after the family moved to Baltimore, at the age of sixty. Mary Gantt's life, however, was crowned by many years, for she lived to be eighty-seven.

Gantt's care for his mother is one of the beautiful parts of his record as boy and man. He was a devoted son to her. From the age of twelve on he lived but two years in his mother's home, and they were within the period after he had completed his education. From the time he first left home until his mother's death, not a single week passed when he was away that he did not write to her. When occasion offered he visited her at her home in Baltimore. These occasions were many and frequent. As soon as his fortunes permitted he provided her with a home where she might be comfortable and happy, and summer after summer he made all the arrangements for a vacation absence for her. He was indeed a dutiful, considerate, loving son.

In a similar way he was a kind and thoughtful brother.

To his daughter he gave the name of his deceased eldest sister, as if to perpetuate her memory. His sister Mary's testimonial is "He was a good brother to me."

The first work which Henry Laurence did to earn money was during the summer of 1873, when he was twelve years old. Through his own initiative he got a job as cash boy in a Baltimore department store, and with much pride felt that he was doing something to help his mother. In the autumn of that same year he entered McDonogh School. That event brought about a real separation of the lad from his home. The school, which operated as a military organization, granted only ten days' furlough per year, and permitted only four visiting days per year for parents. Thus, the separation of the boy from his family for the nearly five years when he was a student at McDonogh School was almost complete.

Chapter II

McDONOGH DAYS

WHEN Gantt was twelve years, six months, and a day old he entered McDonogh School. Then began an expansion and enhancement of his inherited endowments, a development of natural abilities and shaping of inbred character, which had a far-reaching effect upon the thought and action of his mature years. The impact of environment and influence of example each contributed during this formative period. The school gave stimulating surroundings. The days brought experiences that ripened into wisdom. Three personalities, the lives and precepts of men whose common interest was the education of youth, supplied the immediate direction and continuing inspiration. It appears that the intelligent boy owed much to the opportunities that began to present themselves, and that he so eagerly seized upon, from that first day at school.

The fundamental teachings of McDonogh were honor, love of truth, adherence to principle, and self-reliance. Expressions of these qualities in every-day life are truthfulness, honesty, and manliness. These were a part of Gantt's very self. To the teachings which aided in making them his own he responded readily, accepted their meaning fully, and held them as settled convictions throughout his life. Truly, McDonogh did much for the poor, lonely boy.

The day when Gantt entered McDonogh was the first

school day of that institution. It had just been established as a free farm school for poor boys and was set up on a military plan. On the roll he was recorded as number six in a group of twenty-one. A bit of interest attaches to the name by which he was known at McDonogh, "Harry L." This name Harry he also used on his applications for entrance to the university and the technical institute. It seems to have been a boyhood name used by others than his family. In his own home, and among his blood relatives, he was always known as Henry, and Laurence. Among his wife's relatives he was always called John, for the reason that "he looked as if his name ought to be John." To his engineering classmates he was "Duffy." During his professional life and among his associates, in terms of affection he was referred to as "H. L.," "Father," and "The Old Man."

Within a few weeks of the opening of the school Gantt rose to the head of his class and maintained that place of honor continuously for the four and a half years of his preparatory-school life. In the summer of 1874, when he was thirteen years old and one of the younger boys, he was made first sergeant, thus becoming the highest ranking student officer. For the succeeding four years he discharged the duties of that position with fidelity, zeal, and success. However, a student officer, except for the hours of drill, had no more authority and responsibility for discipline and good behavior than any other boy. Gantt always exhibited a willingness to do work, was studious and serious. He resented pranks and horse play as interfering with more important things. He was quick-tempered but fair in his treatment of others. He had some fights when they seemed necessary, but was never intolerant or a bully. On several occasions he

McDONOGH DAYS

took prizes for general excellence in his studies. He won the first prize ever offered at the school for English composition, and an honorary scholarship which permitted him to attend a university.

A reproduction of the record of his standing and scholarship at McDonogh appears on an accompanying page. The classes which he led were, to be sure, small in numbers, but the position of leadership which he early established he maintained throughout the succeeding years at both the university and the technical institute. His honorable discharge came on August 12, 1878, when he had been admitted to Johns Hopkins to enter the class of 1880.

Three great men, each of whom made a contribution to the growing boy, have been referred to. One of them was his benefactor who died before Gantt was born; another was his teacher, leader, and friend at McDonogh; the third was his particular hero. To name them: the first was John McDonogh, the wise, upright, austere, industrious merchant prince, the lonely, idealistic, religiously inclined philanthropist—referred to and caricatured as a miser—the far-seeing, kind-hearted, righteous founder of the school who impressed his life-long vision and purpose upon his executors by writing: "The first, principal, and chief object I have at heart (the object which has actuated and filled my soul from early boyhood with a desire to acquire fortune) is the education of the poor (without the cost of a cent to them). . . ." Colonel William Allan was the second. He was the creative educator, the first headmaster at McDonogh, the one who planned its organization, developed its methods, and made it as its founder wished, "a molder of men." The third was General Thomas J. Jackson ("Stonewall" Jackson), the brilliant Confederate general, main reliance of General Lee, severe disciplinarian, idol of his own troops, whose courageous spirit was selected as one of the few to be immortalized upon the walls of the beautiful Battle Abbey in Richmond, Virginia.

John McDonogh was one of the wealthiest men of his day in New Orleans. He died in 1850, leaving a will which provided that the residue of his property was to be devoted to the establishment of free schools for the education of the poor in the cities of New Orleans and Baltimore. He had been born near the latter city. The estate as he left it was invested in real property. The control of the estate was to be in the hands of six commissioners or agents, three to be appointed by Baltimore and three by New Orleans. They were to manage the property under his instructions and distribute the revenue in accordance with the provisions of the will. Several attacks were made upon the validity of this instrument. The final court judgment in favor of the cities was handed down in June, 1872.

As soon as this decision was rendered the board of trustees appointed by the city of Baltimore proceeded to the establishment of the McDonogh School. The fund in their hands was then slightly over $700,000, besides certain unsold property in Louisiana. A scheme for the school had been worked out and approved, a principal and teachers appointed, and a tract of land of 853 acres with a residence and a group of farm buildings purchased in a location about twelve miles northwest of Baltimore.

By an express mandate of the founder's will, the trustees established a farm school. At the same time it was recognized that the primary objective of the school was education and that farm work was but one means among many to be used in reaching the goal. It was customary to divide the boys for work into squads, each under a leader. Military officers were uniformly selected for this responsibility, which meant that Gantt became early a gang boss, or foreman.

The farm was under the direction of a farmer, who employed five or six laborers to care for the animals and do the hardest part of the work. However, there was plenty for the boys to do. In the summer-time the rising hour was 4:30 A.M. and even in the winter-time only one hour later. They planted, hoed, cut, stacked, and husked corn; stacked the wheat after it was cut and bundled, and later on assisted in tending the threshing-machine when it was threshed. After the hay had been cut they did all the work of curing it, loading it on the wagons, and mowing it away in the barns. They did the gardening, which included raising all the vegetables for the school, and small fruits such as strawberries and raspberries. When there was a surplus of berries the boys assisted in peddling. Of all this work Gantt did his full share, and to the hearty, healthy life of which it was a part, may be attributed in some degree his physical development, broad shoulders, deep chest, and strengthened vitality sufficient at any time to tire out his younger associates. At the age of fifty-four, with no training whatever, he walked seven miles in one hour and forty-seven minutes.

The home which became the first school building at McDonogh, while none too well adapted to its new purpose, had a certain spacious comfort and elegance which might well have had an influence on the boys, most of whom came from homes of meager possessions. In a sketch of the first winter and its accompanying difficulties, written a dozen years after the school opened, John Johnson, Jr., a classmate of Gantt's, and his particular rival in scholarship, gave this vivid account of the inconveniences, discomforts, and hardships that the boys endured:

> The first boys reached the school when winter had fairly set in, and when city boys could hardly be expected to enjoy a transfer to the country. Their first winter at McDonogh was not spent without enduring a good many discomforts. Some of these, such as the inconvenience due to the lack of a play-room for use in bad weather, it was found impossible to remove until a new school-building was put up. The house in which the school was opened had been originally built as a private residence and had been hastily fitted up as a school by adding a frame wing of a rather insubstantial sort, containing a dining-room, recitation-rooms, and a large room made to serve as a study-hall, play-room, and assembly-room for any exercises requiring the whole school to be called together. As it was often imperative that the boys should be kept out of this room at times when many of them were at liberty, they were together sometimes for hours without a comfortable lounging-place. In such times they took refuge from the cold and wet in the cellars containing the boiler that furnished steam to heat the house, or they scattered themselves over the barns and stables and other outhouses.
>
> During that first winter the organization of the school was perfected. The boys were arranged in a military company and required to wear a uniform. The company was officered by the older and more mature heads among them, and these officers were given considerable power and responsibility in the maintenance of discipline in minor matters.

The main studies were mathematics, elements of natural science, drawing, music, English, German, and American and English history. The schedule required seven hours of each school day on these studies from eight o'clock in the morning until one in the afternoon, and from seven until nine in the evening. In the former period both preparation and recitation were included. The latter was for preparation alone. Boys who failed in recitation were required to spend a short time during the afternoon, not more than forty-five minutes, in making up the deficiency. Thus there remained the greater part of each day, and the whole of Saturday, for the farm work which the founder wished to give to his beneficiaries.

Mention has been made that the school was organized as a military unit. In addition to the discipline which this gave to the youngsters there was developed under the leadership of Colonel Allan an honor system which threw the matter of judging and punishing minor offenses upon the boys themselves. This part of the training is often referred to by McDonogh graduates of those days as one of the most important bits of the training that they received under the man whom Gantt at one time referred to as "The Dr. Arnold of America." There were in Colonel Allan a gentleness and strength which made him an ideal executive. One of his boys remembers that when the Colonel sent for an offender he never said, "Did you do thus and so?" but, in grave kindliness, "Why did you do this?" Thus he obviated the danger of a denial on the part of the culprit and, as it were, paved the way for an explanation which was really a confession.

The economic situation of the boys meant that they had little or no money. Partially to overcome this situation, at least to provide them with "currency," there grew up a system of good marks or credits, and poor marks or debits. From Johnson's interesting description of this system the following paragraphs have been adapted:

> Those boys who had offenses charged against them were put to work during play hours at such tasks as chopping wood, cleaning the yard, or weeding the garden. Their more fortunate school-fellows were at play or on the ballground. Peccadilloes, when reported by the officers whose duty it was to note them, were recorded in a book kept for the purpose. They were atoned for by the healthful penance of outdoor labor. The book in which these wrongs were entered for a reckoning was known as the "work-list book." Incurring a penalty was called "getting on the work-list." Each boy's name was written in the work-list book, and after it was put a list of his small sins. These were reckoned according to a prepared tariff of indulgences, as they might be called.
>
> For example, the boys were required to keep the buttons sewed on their pants and coats. If one of the youngsters saw a button fly off he knew that if he did not replace it before the next inspection he would receive one demerit in the work-list book, or, in the slang of the school, he would "get on one day for the button off." By "one day" he would mean one day's work, or as much work as is ordinarily required to be done in the afternoon playtime of one school day—pretty active employment for about one hour and a half. Slovenliness in other respects, as a dirty face or an unbrushed head, brought the same penalty. More serious offenses were more heavily punished by requiring more work to atone for them; the scale of fines ranged up to twenty days, the penalty for going out of bounds.
>
> Value, other than that due to the power of canceling offenses, was given to the credits by various customs and regulations. In the first place they could be exchanged among the boys. In the second place boys deeply in debt could "buy" credits from their better-behaved companions. These purchases were sometimes made with pocket-money, more often with the good

things sent by friends at home. To those boys who had no offenses to answer for, their accumulated credits were a source of profit. Further value was added by the permission to buy holidays with this novel money. A boy who had laid up fifty credits was allowed to visit his friends on Saturday, in return for which privilege twenty-five credits were removed from his account. Additions were also made to the length of the boys' summer holiday for a sufficient number of these good marks.

The farm, having an area of more than a square mile, afforded numerous opportunities for sports and play. The common games of American boys were highly popular. In addition there were certain sports somewhat peculiar to McDonogh. Gantt apparently did not enter into any of the organized games, but he was fond of the opportunities for sport offered by the woodlands and swamps. In the creeks were fish, in the swamps muskrats, and in the woods rabbits and nuts. Catching fish, trapping small game, and gathering black walnuts were the favorites. Swimming facilities were ample. A pleasant little creek furnished plenty of places suitable for a plunge, and in winter skating was possible on ponds from which the ice was cut.

Indication has already been given of the rigidity of the discipline at McDonogh, which in Gantt's time was far more severe than in most preparatory schools. At the outset, appointment to the school was for a period of but one year. Reappointment had to be won and depended principally upon the boy's marks. Upon these marks depended the length of holiday the boy was permitted to spend at home during the summer vacation period. They also added much weight in the award of the coveted prize scholarships. These gave the successful competitor an extra year of tuition and maintenance.

Under the influence of this system the boys felt from the first that they had to win their places, that their chances for success depended upon their own ability, industry, and character. This idea of individual responsibility was also applied in minor matters. Every boy was held to strict account for all school property intrusted to him, including books, tools, and clothing. All of these articles were inspected periodically, and punishment was meted out for damage. If a boy had carelessly torn his coat or burned his boots before the date for issuing supplies, he could get no more from the school and must go ragged and ill shod unless friends or family came to his relief. The purpose of this severity, in the words of Colonel Allan, was to establish a "steady and strict discipline, the object of which is to encourage and develop industry and self-reliance, as well as truthfulness, honesty, and manliness."

H. R. Fardwell, who was one of the younger boys during the early years, recalls the wholesome respect everyone felt for Gantt, their head officer: "Gantt gave promise of his career early, and was the outstanding youngster in the crowd. He was serious-minded and very deliberate in everything he did. So far as my recollection serves me, he was not particularly companionable with any of the boys; however, all of us had the greatest respect for him and for his talents. He had much to do with forming the early ideals with which McDonogh history reeks. He was the soul of honor and a great favorite of Colonel Allan's."

Mention has been made of the influence upon Gantt of three great personalities which touched him during McDonogh days. Each deserves a brief characterization to show the contribution to his development.

McDONOGH DAYS

John McDonogh's ideals, in addition to governing many of the decisions and acts of the trustees and faculty of the school, were presented to the boys for emulation. Each year a celebration in his memory was held on November 21st, Founder's Day. As this was so near Thanksgiving the commemoration partook of that nature, with turkey and ice cream in abundance, and trustees and visitors joining in the festivity.

One tender touch reminds each generation of the boys of the school, of the human longing to be remembered that yearned beneath the grim exterior of John McDonogh. He requested that once each year the boys of the School should lay flowers upon his grave. When May brings its bright blossoms to embroider McDonogh woods and meadows, the pilgrimage is made, and appropriate words and music make up the simple tribute. The McDonogh poet laureate, Margaret J. Preston, singer of the South in her dark hours, sister-in-law of Stonewall Jackson, and stepmother of Colonel Allan's wife, wrote the ode used for many years. One verse reads:

> None call him father who might lay
> Love's tribute on his sleeping breast,
> But we who heed his last request
> Are we not all his sons today?

The character and career of this man of strong personality and powerful imagination are somewhat revealed in a set of rules which he drew up for his own conduct when only about twenty-four years of age:

Remember always that labor is one of the conditions of our existence.

Time is gold: throw not one minute away, but place each one to account.

Do unto all men as you would be done by.

Never put off till tomorrow what you can do today.

Never bid another do what you can do yourself.

Never covet what is not your own.

Never think any matter so trivial as not to deserve notice.

Never give out that which does not first come in.

Never spend but to produce.

Let the greatest order regulate the transactions of your life.

Study in the course of your life to do the greatest possible amount of good.

Deprive yourself of nothing necessary to your comfort, but live in an honorable simplicity and frugality.

Labor, then, to the last moment of your existence.

Pursue strictly the above rules, and the Divine blessing and riches of every kind will flow upon you to your heart's content; but, first of all, remember that the first and great study of your life should be to tend by all the means in your power to the honor and glory of the Divine Creator.

The degree to which these rules governed John McDonogh's life is indicated by his adherence to the next to the last, which reads: "Labor, then, to the last moment of your existence." When seventy years of age he devoted eighteen hours a day to work. On a certain day, when nearly seventy-one, he spent his working-hours as usual in the city, returning to his home late in the afternoon. Two days later he was gone. His energies he devoted to winning wealth and achieving the ambition of owning a ring of land entirely around the city of New Orleans. Commenting enthusiastically on this fact to one of his friends a few years before his death, he said:

"Congratulate me, my friend, I have achieved the great-

McDONOGH DAYS

est victory of my life. I have drawn my lines around the city [New Orleans] and now entirely embrace it in my arms—all for the glory of God and the good of my race." His wealth, perhaps the most extensive landed property then owned by a single individual in the United States, he consecrated to charity, looking upon himself merely as a steward intrusted with means to be devoted to the welfare of his fellow men.

Gantt's life showed the guidance of many of John McDonogh's rules and much of his idealism.

When the trustees of the McDonogh School instituted a search for their first principal they turned for counsel and suggestion to General Robert E. Lee, then president of Washington College. Lee recommended Colonel William Allan, a member of his faculty. Allan was a Virginian by birth and Scotsman by ancestry. His parents, coveting an education for their son, sent him to the University of Virginia, where he took his Master's degree in what was then known as "mixed mathematics." For about two years he taught at a school in Albemarle County, Virginia. When the Civil War broke out he presented himself for examination for a commission in the Confederate army. He passed *primus inter pares* and was made Chief of Ordnance in Jackson's corps of the Army of Northern Virginia. In this important and trying position he gave a most creditable account of himself until the close of the war in 1865.

Like his great commander, Lee, Colonel Allan never allowed himself a bitter or unkind word concerning the fratricidal strife of the Civil War. But the cause of the Confederacy was ever cherished in the deepest recesses of his heart. He devoted his entire leisure to preparing fair and accurate war memorials which are now highly-prized source books. *Jackson's Valley Campaign* in especial degree won well-merited commendation.

In an appraisal and appreciation of Colonel Allan after his death, the Honorable John Randolph Tucker, a distinguished jurist and Congressman of Virginia, paid tribute to his love for truth, self-effacement, and devotion to principle. To quote Tucker: "Colonel Allan was a man who dealt in abstract truth, and then wielded it for practical advantage. He knew and felt that at last the thinker strikes the keynote which directs the orchestra of human voices in the resounding chorus of the world's progress and civilization." Love of truth was one of Gantt's outstanding characteristics.

Tucker emphasized the self-effacement of Colonel Allan in this paragraph:

> He gauged his powers justly and did not over-estimate them. He was conscious of possessing them. He had self-esteem without a particle of self-conceit, and while he valued the esteem of others, he was without pride or vanity. He was singularly simple in self-estimate. He knew his capacity for labor and for the practical administration of affairs. He valued power not for himself but as it gave opportunity for good. He was self-reliant without presumption, and modest without self-distrust. He knew he was honestly purposed to help poor boys to be self-reliant, brave and noble men; to be rich in themselves though poor in estate. He had been all that himself.

The attribute of modesty was a significant element in Gantt's make-up.

Tucker also particularly stressed Colonel Allan's adherence to principle:

> His convictions were strong and these he enforced with self-respect, but without arrogance or unpleasant self-assertion. He

McDONOGH DAYS

sought more to convince the judgment than to compel the will. When his cogent reasons failed to convince, his nature never sulked in his tent, when the battle was imminent; but yielding to what was not in accord with his own opinion he magnanimously joined in the promotion of the accepted policy with a generous purpose for the public good and to achieve success for the common cause. But, if essential principle was involved, he would not yield this to the dictates of expediency. Gentle in manner, his conscientious conviction made him sturdily resist, and with intrepid zeal, any policy which violated moral principle.

Mrs. Preston's poem, written for the unveiling of Valentine's bust of Colonel Allan, a gift of the alumni, has the same thought, for principles and convictions are "living lines":

> No Raphael's canvas did he use
>   Some high ideal thought to follow,
> No Parian marble did he choose
>   From which to sculpture his Apollo.
> But purer far his lofty thought,
>   And higher yet his brave endeavor—
> To grave on boyhood's plastic heart,
>   Those living lines that last forever!

In these regards Gantt was strikingly like him. Unyielding, uncompromising, he always searched for the principle involved, and once it was discovered could not be swerved from its application to the very end.

As to the influence of the picturesque personality of "Stonewall" Jackson, the record of the life and career of that brilliant soldier, skilled strategist, and master tactician so captured the imagination of Gantt that the General became his especial hero throughout his entire life. Indeed, there are many points of similarity in the character of these two men, the one who fought battles of war, and the other who fought battles of industry.

Jackson was virile, devoted to duty, possessed of a high sense of self-reliance. He was a leader of proved daring, enduring courage, the courage that glories in swift action and the excitement of the charge. He was a man of many sides and sharp contrasts. At home he was the genial, happy, fun-loving host and companion, delighting in stories, repartee, and harmless practical jokes. In the community he was the exceedingly charitable, zealous, praying churchman. Toward all he had the "delicacy and the tenderness which are the rarest and most beautiful ornament of the strong." He was an ardent lover of nature and the out-of-doors. The glories of the autumnal foliage, the songs of the birds, the splendors of the sunsets, were sources of unfailing pleasure. These beauties were especially close to him during the years when he lived at Lexington as Professor of Artillery Tactics and Natural Philosophy at the Virginia Military Institute. Toward the eastward from his home were the huge masses and lofty peaks of the Blue Ridge Mountains. Toward the northward were rolling hills green with meadow and forest sloping gently to lower levels. Winding through the valley was the lovely river which the Indians called the "bright daughter of the stars," the Shenandoah. On moonlight nights this queenly beautiful Valley of Virginia seemed to shine with brightness unequaled elsewhere. It was General Jackson's custom, weather permitting, to take his wife for an evening walk or drive that together they might enjoy these beauties.

One can imagine how Colonel Allan must have talked of his old chief, telling the boys of the outstanding, vigorous

McDONOGH DAYS

character of the great warrior, of his military genius, of his strategy and tactics, of his methods both in attack and in retreat, of his victories and of his untimely death at the hands of his own men. One can also imagine how young Gantt listened with rapt attention to this presentation and how he must have been impressed by facts like these which showed, on the one hand, fairness, and on the other adherence to principle: On one occasion, when Jackson was a professor at the Virginia Military Institute, he walked two miles through the snow after ten o'clock one stormy night in order to apologize to a boy whom he had unjustly reprimanded in the classroom during the day. During the Civil War, when he was one of Lee's most trusted generals, he was ordered into Richmond for a military conference. He rode all day Saturday until the stroke of midnight, then stopped, put up his horse, and went to bed. During Sunday he attended church, as was his custom. Exactly at the stroke of midnight Sunday he swung himself again into the saddle and pushed on for Richmond.

Whatever may have been the introduction that Colonel Allan gave his former leader to young Gantt, it was sufficient to arouse an enduring interest. Gantt read everything he could find in print written by or about "Stonewall" Jackson. The book he read the most was Henderson's *Stonewall Jackson and the American Civil War*. Gantt was like Jackson in his fearlessness, ability to make quick decisions, almost complete lack of self-interest, and love of the beautiful.

Truly McDonogh was a "molder of men" under Colonel Allan in the days when Gantt was a pupil. The influence of those days, and the examples of these three leaders, seem to have shaped much of the purpose and thinking of Gantt's mature years.

He returned to McDonogh again and again, as a member of the faculty, as an overnight guest, to attend celebrations, to speak on special occasions. He made numerous gifts to the library, kept in touch with the faculty, and watched with interested care the development of the school where he spent nearly eleven years of his life in residence.

CHAPTER III

JOHNS HOPKINS AND STEVENS

AFTER Gantt had attended McDonogh for about four years he began seriously to plan for further education. In this purpose he had both the wise counsel and warm encouragement of Colonel Allan. Two personal problems had to be solved—those of tuition and living expenses—before the door of opportunity would be open wide to the ambitious youth. The first of these needs was met by a scholarship granted by the president of the Johns Hopkins University; the second was satisfied by the honorary scholarship won at McDonogh. Thus the university gave his tuition and the school paid his living expenses. This arrangement permitted him to continue to live at McDonogh while attending Hopkins.

His application for admission to Johns Hopkins was dated June 10, 1878. One of the questions requested the candidate to state somewhat fully the purpose which he had in view in applying for admission. Gantt replied: "I wish as liberal an education as I can get and propose to take one of the scientific courses with reference to fitting myself to teach and perhaps subsequently to enter a profession." In this statement he was prophetic.

At seventeen years of age the average American youth has little purpose in life and has not made up his mind as to what work he wants to do, what occupation, business, or profession he wants to enter. Ask him to tell you where he wants to be at thirty-five, what place in life he wants to occupy, what his shining vision is for himself, and he will look at you blankly, without a single idea that he can express in answer to your query. Gantt was an outstanding exception in this age group, for at seventeen he did have a purpose, he did know what he wanted to do. What is more, he followed out his purpose, for he did teach and subsequently entered a profession.

Two comments from professors who examined him for admission to Johns Hopkins show something of the bent of his mind and a little as to his preparation. In mathematics the report states: "Passed a matriculation examination in mathematics *very well*." Another records, "Henry L. Gantt is well prepared in German to enter the advanced class. In French he has not read much but is tolerable in the grammar."

Gantt entered Johns Hopkins in the autumn of 1878. The university was young, having been founded two years earlier. The classes were small. The subjects which he took during the two years' work which won him his degree included French, German, Advanced Mathematics, Physics, Latin, Logic, and the History of Philosophy. His scholastic record was similar to the one at McDonogh. His marks were high and standing "good" to "excellent." In seven of his courses he was first in his class. This substantial achievement is all the more remarkable, for it was built upon imperfect preparation.

As he commuted daily from McDonogh, his contact with his classmates was restricted. He did not participate in even the limited social affairs held outside of the classrooms. There were two fraternities which were organized at about this time, but he joined neither. He was not musical, so did not belong to the musical club, nor did he participate in the activities of the debating society. Nevertheless, his personality made an impression upon his associates that persisted, even though they saw little or none of him during his professional career. Gantt was not one to be easily and quickly forgotten. He was likable, popular, and respected.

He was fond of commenting on his lack of musical ability. "At one time," so he said, "I took singing-lessons. When they were over I asked for my bill. The teacher said there would be no charge 'if you will not tell anyone who taught you.'"

One of his classmates of those days says of him: "Gantt was a good, steady student and we all liked him." Another, Reverend Carl E. Grammer, comments upon the sturdy, independent and yet modest character revealed in his college days, upon his aim in life, which was to serve, and continues: "My earliest recollection of Gantt is a report that one of the boys at the McDonogh School, outside of Baltimore, had so distinguished himself by his proficiency that President Gilman of the Johns Hopkins University had procured him a scholarship in the Collegiate Department of the University. At that time a certain amount of Latin was required for matriculation, and Gantt soon appeared in the chapel before the morning prayers, which President Gilman used to conduct. Gantt came early and for fifteen minutes before the service Professor Charles D. Morris, the Collegiate Professor of Latin and Greek, used to hear him recite his Latin lesson. I vividly recall Gantt's appearance. He had the solidity and modest dignity, even then, that was one of his marked characteristics."

Another classmate comments upon the seriousness of the

group at Johns Hopkins during the years when Gantt was an undergraduate, saying that they all were influenced to do good work by the presence of a much larger number of fellows and graduate students. Professor Harry Fielding Reid says of him, "He was a good student, especially in mathematics." Thomas Pettigrew remembers him as "a nice, cheerful, good-looking youngster ready to laugh at you and with you, and therefore popular with both students and teachers. He had a liveliness and wide-awakeness of manner which was very attractive."

Gantt was graduated from Johns Hopkins in June, 1880, but did not receive his diploma until the tenth anniversary of the founding of the university, April 26, 1886, due to the practice which prevailed at that time. The quality of his work is indicated by these facts. He acquitted himself with distinction, completing in two years the courses required for a Bachelor's degree, and in addition won a university scholarship presented at the time of his graduation. He declined, however, to accept this honor, preferring, instead, to follow his inclination to teach.

The opening to become a teacher was from McDonogh. For three years, from the 1st of August, 1880, Gantt was a member of the faculty of that school. Thus he rounded out nine and one-half years of continuous residence. He taught the natural sciences and mechanics, and each year took charge of the fifth class, the boys just entering. In that position, one of his responsibilities was to form in the youngsters such habits of industry as were necessary to make sure that they might continue happily and successfully at McDonogh.

One who was neither student nor faculty member—Colonel Allan's daughter Janet, later Mrs. William J. Bryan—

JOHNS HOPKINS AND STEVENS

but one who saw much of Gantt during these teaching years, recalls him as "Always a keen, sharp-tongued chap, able and far-visioned, but not too affable or suave."

These three years at McDonogh did not complete Gantt's teaching-record, however, for in June, 1886, he returned to McDonogh after having completed his technical education and having worked two years in industry. The occasion of his return was the receipt of a bequest of $80,000 by the school to be used in establishing manual training. A beginning was made in wood-working under Solon Arnold, a graduate of the United States Naval Academy. In June, 1886, Arnold was ordered to join his ship, thus leaving a vacancy on the faculty which Gantt was asked to fill. This post he held for a little more than a year, leaving July 6, 1887, to go to the Midvale Steel Works. During that time he developed the Barnum Shops to teach not only wood-working but also iron-working and mechanical drawing.

To turn back to the closing months of his first period of teaching: Early in 1883 Gantt began to cast about as to his next move in life, feeling strongly the need of more education. On March 12th he wrote to Dr. D. C. Gilman, president of Johns Hopkins, presenting his problem in this way:

> Ever since leaving the University I have been looking forward to the time when I should return, and for some time past have been thinking of next fall.
>
> Now I wish to lay before you in this letter my circumstances briefly and to ask you to give me a few minutes some Saturday when I can talk the subject over fully with you and get your advice.
>
> Is it possible that I could get such a place at the University as, with the $500 I now have, would enable me to get along for three years? Or even two years? For two years' work would be

better than none. If I remain here another year I shall scarcely be better off then than now, and the difficulty about going to the University will be the same.

My object in going to the University is to study Physics as major and Mathematics as a minor subject.

President Gilman, with penetrating insight into Gantt's abilities, and keen judgment as to the nature of further education which would be most helpful to him, combined with a full knowledge of the personality and program of a fellow-pioneer educator, President Henry Morton of Stevens Institute of Technology, advised him to go to that institution instead of returning to Hopkins for graduate work.

On June 8th of the same year Gantt wrote again to Dr. Gilman saying that he had made satisfactory arrangements and was to enter Stevens. He included this comment: "The expenses will be considerably more than I anticipated, but I have had instructions from two of the faculty that I can probably get some work to do. I shall go up there about the 1st of August, thereby putting in two extra months."

A letter of almost a year later, June 4, 1884, is in the nature of a report to Dr. Gilman in regard to the school year which was just closing, and expresses appreciation for his "kind and valued direction."

My course at Stevens Institute being about ended, I have thought that a short account of how I have followed your advice would not be uninteresting to you. Finding by my visit here last spring that a little work during the summer would enable me to enter the senior class and take the most important part of the course devoted specially to Engineering, I concluded to do so.

On entering the class I showed Professor Mayer some of the reports that I had handed Dr. Hastings and was excused from laboratory work in general Physics, and given an opportunity to put the time thus gained on Electricity.

As most of our class had taken the regular course here, they had done a considerable amount of work in the shops of the Institute while I had done none. I accordingly made an effort to do some of this work and succeeded in spending about three hours a day in the machine shop, until nearly Christmas. During the winter term I did not have so much time to spare from my studies, but succeeded in doing a little pipe-fitting, and giving some assistance in a test of two powerful Weston Dynamos which was made here.

In March and April the class made a tour of inspection to Philadelphia and the neighborhood, and then to Boston, Providence, and other places in that region. These trips served very materially to open our eyes to what was going on in the manufacturing world.

On our return from the last of these trips there was nothing to be done but write our theses.

An engine in New York, using vapor of bisulphide of carbon instead of steam, has been attracting considerable attention, and Professor Thurston suggested that, for my thesis, I should discuss theoretically the working of several vapors, if used in place of steam, and compare them with steam. Two of us, D. H. Maury, a graduate of the Virginia Military Institute, and I, have been working together on this subject and expect to finish it this week. As Maury has been here only two years, he has not had much shop work, so we worked on our thesis in the afternoons and evenings, and spent the mornings in the shops, averaging over three hours a day for the last seven weeks. As there were only five or six students taking this course, the instructor had a great deal of time to devote to each one, and our time was very profitably spent.

The scholastic record which Gantt had made at McDonogh and Hopkins was repeated at Stevens. Minutes of the faculty show that he was admitted conditionally to the senior class, and that he then proceeded to earn average marks of 97.2 for the first term, and 89.7 for the second.

In Gantt's letter to Dr. Gilman, which has been previously quoted, he mentions that Professor Thurston (Dr. Robert H. Thurston) had suggested as a thesis a theoretical discussion of the action of several vapors as working fluids in an engine. This thought was acceptable and the joint thesis of Henry L. Gantt and D. H. Maury was devoted to the following:

> History. A brief account of some of the more remarkable inventions in vapor engines.
>
> Problem I. A theoretical discussion of the behavior of the vapors of water, alcohol, bisulphide of carbon, and chloroform when worked in an engine, between the limits of pressure common in the steam-engine of today.
>
> Problem II. A discussion of the behavior of the same vapors when working with the same initial pressures as in Problem I, but with final pressures regulated by the back pressures, and differing from them by a nearly constant quantity.
>
> Problem III. A discussion of the behavior of these vapors as worked between the limits of temperature common in the modern steam engine.
>
> Problem IV. A similar discussion with the initial pressures of Problem III, and the final pressures of Problem II.
>
> Problem V. A discussion of the action of vapors worked in Carnot's cycle.

This particular thesis won far more recognition than the usual undergraduate effort, for it was published in *Van Nostrand's Engineering Magazine* in November, 1884, thus becoming the first professional contribution from Stevens '84.

Gantt's graduation from Stevens was on June 12th, with the degree of Mechanical Engineer.

Gantt, as memories of fifty years bring him back to his

classmates at Stevens, was dignified, purposeful, attentive to his duties, or as Harvey F. Mitchell puts it: "a quiet, self-contained, industrious man, a good student, not represented on the athletic field or in the Glee Club." He did, however, join the Beta Theta Pi fraternity.

The reaction to his lack of interest in athletics is still strong. C. W. Thomas expresses a bit of feeling in this way: "Gantt was not interested in athletics. I was on the football team, and although he was heavier than I he would not come out on it. It seemed to me he acted a little uppish because of his being from the South."

As to Gantt's own appraisal of his year at Stevens we have an additional flash. Speaking of education one day to Charles B. Going, he said: "Most of the stuff I learned at college I never used, but, thanks to one man, I got something out of it that was worth more than all the rest. That was a way of seeing things. Professor —— was not a very great teacher. He was not a great mathematician, for I caught him in mistakes. But he gave me a point of view that changed everything in life, and whatever I have been able to do is due to that." Going is not entirely certain, but believes that this reference was to Professor Thurston.

Other events also made their impression, for Gantt was more mature and experienced in life than possibly any other man in the class. The influence of Johns Hopkins and the years of teaching at McDonogh had developed a point of view and strengthened a purpose that shaped the acts of the years of his strenuous life. In fact, he seems to have begun to practice at McDonogh, while teaching, the very principles that he so vigorously supported and applied in his professional work.

With Gantt in his special summer course at Stevens was William S. Aldrich. They worked shoulder to shoulder to make up deficiencies in industrial and engineering chemistry. Between them was a close sympathy, for each was as a stranger in a strange land among the seniors who in a few months were to be graduated as the class of '84. Aldrich's preceding education had been at the United States Naval Academy at Annapolis. With the feeling of a warm friendship Aldrich writes of Gantt:

> President Gilman of Johns Hopkins University had evidently made a strong impression upon him. Gantt's years under that great educator started within him the evolutionary growth of a scientific spirit and respect for the scientific method. And, further, respect for and an acceptance of intellectual honesty and integrity as necessary qualifications for scientific development. That is, Gantt's intellectual honesty and integrity, built upon his own inborn traits, became a specific life characteristic, definitely in its directed development a heritage from Hopkins.
>
> It further appears that Gantt was not only well prepared as to education and training for his specific life's work, but his whole self was centered in, and bound up in, the purpose to improve and better industry. His will, intellect, and affection, the whole soul of him as it seemed, was seeking through his work to give expression like the creative imagination of an artist to his inborn humanitarianism.
>
> Gantt was perhaps quite unconsciously pragmatic in that he believed that what worked at McDonogh would also work in the larger field and intense life of industry and its effective management. Two features he explained to me, speaking of his experiences at McDonogh, were:
>
> >   Teach them—
> >   To teach themselves
> >   To govern themselves

JOHNS HOPKINS AND STEVENS 45

In this respect it seems his patron saints were: Herbert Spencer, "educate men to be self-governing creatures," and Thomas Jefferson, "the best governed are the least governed."

His fundamentals of action, basic humanity and ideal humanitarianism were apparently early accepted and were surely reasonably well in hand by the time of his entering Stevens. It would be difficult now to state just what determined and decided this early trend, preference, and predilection. Most probably it was a trait both inborn and inbred. Thus his life's course was set and point of departure determined by the time he launched out from Stevens. He had evidently applied, evolved, and developed what he had to build with and upon, satisfactory to himself and his environment, in his three years of multitudinous teaching contacts and experiences at McDonogh.

These attributes of personality which he brought with him certainly worked at Stevens. Although Gantt came a stranger among strangers, in a strange land and environment, his kindly interest and sympathetic disposition were at once apparent. In a certain sense he was "all things to all men," students then and workingmen afterward. Whence we find in him a disciplined humanitarian in social-ethical-economic relations, sympathetic to his fellow men, aiming to provide for their common good in these three aspects and to promote their general and individual welfare.

Probably the nickname we all had for him during the senior year's contacts at Stevens covers the multitude of the impressions which he made, "Duffy Gantt."

Chapter IV

HOMES AND FAMILY

Twelve years passed after the close of his teaching at McDonogh before Gantt established a home of his own. Even then his life, like that of most consulting engineers, was the goings and comings of a transient who spends days at work and nights in traveling.

Among his friends he had been regarded for years as a "confirmed bachelor," one whose creative power was given to his work, whose energies were devoted to his employers' and clients' interests, whose constructive thinking was being directed more and more into philosophical channels, such as the problems of industrial organization and relationships between employers and employees. Across the situation fostered by these beliefs there one day came, with the suddenness of a lightning flash, the news that Gantt was engaged to be married to Miss Mary Eliza Snow, of Fitchburg, Massachusetts. His numerous friends hastened to send good wishes. As one and another had the privilege of meeting his *fiancée*, their congratulations became unusually warm and increasingly cordial. Gantt had picked a lovely young woman, three years his junior. In his breezy way he referred to her among his close friends as "the victim."

At the time of her engagement Miss Snow was tall, slender, black-haired, and brown-eyed. Though a girl born to New England of Colonial *Mayflower* ancestry traced back to

John Alden, she was also a daughter of the South. Her mother was a South Carolinian. In a very real way she was a mixture of the two sections. She had the warmth, charm, and grace of the refined Southern woman, to which were added the energy and fire of the cultured woman of New England. She came from a well-to-do home. Her father was a lawyer and man-of-affairs, at one time a State Senator for Massachusetts. In girlhood her health was none too secure. For this reason her education was largely from private tutors. During a five years' residence in Germany and France she studied languages and the piano.

In many ways Mary Snow was the antithesis, or complement, of Gantt. She was gracious and tactful, where he was somewhat brusque. She was exceedingly social and fond of entertaining, while he liked just a few understanding friends around him. She was an accomplished pianist, while he disliked music, even her playing. She was a very good bridge-player, while he did not play at all. She had remarkable grace of manner, sincere withal, while he was apt to "rub people the wrong way." She was extremely well-read keeping up to date with current literature, both books and magazines. He read none of this, devoting himself exclusively to technical and historical subjects.

The marriage did not take place for more than a year after the announcement of the engagement. But on November 29, 1899, Henry Laurence Gantt and Mary Eliza Snow were married in Christ's Church, Fitchburg, by the bride's brother-in-law, Dr. Francis A. Brown.

The changes in location, imperative in the life of a consulting engineer, caused frequent moves in the home that they established: First at South Bethlehem, Pennsylvania, then northward to Schenectady, New York, then on to Providence, Pawtuxet, and Pawtucket, Rhode Island, still later to an apartment hotel on Fifty-seventh Street in New York City, and last of all to what was perhaps the real home of the family, in Montclair, New Jersey. So numerous and disturbing to home life were these moves that Mary once said: "If it were not for Peggy's having a stepmother, I would rather die than move again."

Within a year after they were married a daughter, Margaret Heighe, or "Peggy," as everyone came to know her, was born. At the moment of her birth, Mrs. Gantt's condition was such that the attention of everyone was centered upon her. So the new-born baby was rolled up in a covering and laid aside. Gantt, feeling she was neglected, and with utter matter-of-factness, proceeded to heat newspapers and wrap them around blanket and baby. At the same time he was so solicitous and anxious over Mary that the nurse and his sister-in-law invented errand after errand to get him out of the house and out of the way.

As Peggy began to grow and develop, Gantt adored her. As a little girl with big brown eyes and long brown curls she was the idol of his heart and the joy of her mother. Lillian Gilbreth writes of that period: "I have never known a child more adored; her mother especially gave her life to seeing that Peggy had everything that a child could have—companionship, understanding, complete devotion."

The occasion of the move to New Jersey was this: One day while lunching with one of his friends, Charles Whiting Baker, Gantt happened to mention that he was tired of hotel life and was looking for a place in the country. He said his little girl had exclaimed, on visiting in the home of a friend,

"Oh, see the round pie!" Baker suggested Montclair, one of the pleasant suburbs on the eastern slope of the first Orange Mountain in northern New Jersey. Here Gantt found the home he was looking for at 18 Hoburg Place. It was named "Oakleigh," from its trees.

The residence was a beautiful old house with a sloping lawn in front and a grove of large trees behind. At the rear of the entrance hall there was a pipe organ, and at the left, as one entered, a huge living-room with an immense fireplace midway of one side. Mary Gantt filled it to her heart's delight with antique furniture, much of which had peculiar sentimental value because of personal associations. Many of the pieces were heirlooms of the Snow family. At "Oakleigh" she was very happy because of the history and traditions of the community, the house and its furnishings, and her wide circle of friends. She was also busily and pleasantly engaged in church affairs and with many social activities.

To Gantt the most likable spot of all was the great living-room. Here, again and again, at nearly every weekend, and often in between when he was at home, he would gather friends around him to discuss the pressing problems of industry and business. Here he lived during the trying years of the World War. Here he thought and mused, distilling wisdom from a life-long, ripened experience and evolving those philosophic concepts which, although they have had little impact upon American industrial life up to the present time, are the great heritage which he left to all who have followed and will follow him.

He was also fond of the surroundings of the house. Many an hour he walked back and forth across the lawn, in step with some friend to whom he was earnestly explaining some of his thinking. Often he would pause, glance searchingly at his companion, and ask this favorite question, "Do you get me?" All the time he was seeking for comment and suggestion, the impact of another mind on the problem that was holding his attention. During these leisurely, reflective saunters, Sam, a handsome white, blue, and black Llewellyn setter, was always by his side.

Gantt was extremely fond of hunting-dogs, but had a feeling of contempt for most others. Roy, a pointer, he had raised from a puppy and trained himself. In Pawtucket, Roy was Peggy's much loved, gentlemanly playfellow. Sam was bought in the South and brought back to Montclair. When Gantt was at home Sam was always near by, lying by the fireplace in the living-room or near his master if he were elsewhere about the house and grounds. A special piece of family equipment was a shipping-crate in which Sam was sent by express to join his master on hunting trips. With Sam's aid Gantt brought home many a bag of quail and duck.

One of these hunting excursions yields this story. One evening after a very successful day Gantt brought home an excellent bag of quail. Mrs. Gantt cooked them with especial care and the next morning ten or a dozen were packed for luncheon for Gantt and one of his friends, who were going hunting. They were driven out by a colored boy whom they instructed to drive along the road to a certain point, where they would meet the team after they had hunted across an intervening field. They arrived at the rendezvous about luncheon-time to discover that the hungry negro had eaten practically all of the quail. The vigor and directness with which H. L. expressed his opinion of the

Gantt, Peggy and Roy

Mary Eliza Snow at the Time of Her Engagement

darky's greed is easily imagined by all who knew him and had heard his explosive comments.

Both Laurence and Mary Gantt were devoted not only to Peggy, but to other children, almost all children, so it seemed. They were particularly fond of visiting the Gilbreths and joining in the home-doings of that large, lively, enthusiastic family. On such occasions Gantt had a habit of looking at the young folks over his glasses with a genial, engaging smile, genuinely interested in whatever they were doing or planning to do. He never talked down to a child, he always talked with him. Because of this trait children loved him. He was just as natural and casual with one of them as with any grown person.

On one occasion Gantt had his secretary contact with some of the welfare organizations in New York City with the thought of adopting a boy to "raise in that big house with Peggy," but nothing came of this bit of exploration. During the World War he suggested adopting a couple of Belgian orphans, but Mary objected.

Because of his never-failing sense of humor and the proportion and fitness of things, and because of his absences from home, Gantt left most of the upbringing of Peggy to her mother. In this he was following a principle of his own work. The management of the home and the activities within it properly belonged to his wife. Mary accepted this situation with eagerness. She adored Peggy, was very demonstrative with her, looked after every detail of her life, planned her education, and hated to see her grow up. Perhaps because of this situation, perhaps also because of the kind of thinking and studying he was doing during the last few years of his life, there did not develop the close relationship between father and daughter which sometimes exists. However, the impress of his personality and heritage upon her was strong, and no one who knew "H.L." could meet her without realizing that Peggy was "her father's daughter."

The home life was very happy. The marriage seemed particularly fortunate for both. Gantt brought to it in a large measure the masculine qualities of intellectual power, strength, robustness, courage, aggressiveness, creative energy, and stern justice tempered by generosity and tenderness. Mary brought the feminine traits of patience, kindness, sympathy, self-sacrifice, sensitiveness, to every situation, and domesticity, and withal strength of character. Thus they were very congenial. To this happy situation Mary contributed much through her flexibility and adaptability. She had plenty of well-formed opinions, but always kept them in good balance.

The family life was simple. Guests were always welcome. There were lots of company. Mary was fond of parties, of entertaining, and of going out, but withal she was a real homemaker. In addition to her own circle of acquaintances and intimates she entertained her husband's business friends and their wives; in fact, was on terms of social intimacy with them. Gantt was more quiet, deeply fond of his home from which he was so often absent, and which provided an ideal setting for his philosophic musings and discussions. Mary was entirely sympathetic with his work, proud of him and his achievements. Gantt, on his part, was equally proud of Mary's social successes. Their friends often said of them, "A wonderful fit in the family relation."

HOMES AND FAMILY

Mary's interest and appreciation of her husband's contribution to the business and industrial problems, to which his energies were devoted, is shown by this excerpt from a letter written April 17, 1920: "My greatest interest in life, next to Peggy, is seeing Mr. Gantt's work spread and develop. Every day I feel more and more strongly that the principles which he preached and which governed his life are the only solution of this terrible condition we are facing."

The respect that each had for the other is shown by their attitudes toward the church. Mary was active in church affairs and regularly attended services. Gantt had a rather poor opinion of churches and preachers. This feeling may have had its rise in his boyhood days. His mother was deeply pious, a high-church woman, and extremely ritualistic. Gantt in his very make-up could not endure what to him were senseless repetitions. Soon after they were married, at Mary's request he did accompany her to a church service. Two things happened. During a solemn moment he sneezed and thus attracted startled attention from other worshipers. Gantt's sneeze was always loud and explosive. At the close of the service he expressed an opinion of the preacher, in a voice audible to many who were around them, to the effect that he was "both a knave and a fool." After that experience he seldom was asked to go to church.

His fondness for experimenting was carried even to the home dining-table. On Sunday evening, particularly, he was fond of cooking or, as members of his family said, "concocting messes." None of the results of these efforts is remembered with approval. It is quite certain that on the following Monday morning the maids found both kitchen and pantry in need of a reorganization.

Gantt's keen sense of justice sometimes led him into situations of embarrassment to Mary and Peggy and to his associates and friends. If the service he received in a restaurant, or when traveling, was not to his liking, he did not hesitate to say so in no uncertain terms. But one had to know and understand the man to realize the source of these flashes of resentment.

In spite of his absorption in his business, Gantt was no recluse. He was fond of jokes, but was not the equal of Mary in the ability to tell of incidents and relate anecdotes. On numerous occasions he accused his wife of saving up all her adventures and happenings, to dress up the funny side of them, and to spring them on him at the first opportunity. Anyone who had contact with him during those times, who had the privilege of talking human-interest matters with him, will always remember him as never sarcastic, unkind, or bitter, but always genial, interested, and sympathetic to everyone.

An incident which shows his gay, good humor, an occurrence on a street in Philadelphia, is recalled by his classmate, Dr. Carl E. Grammer: "I was hurrying to some important engagement, absorbed in thought, when I found my way blocked by some one coming from the opposite direction. Seeking to pass to the right, the way was still blocked; turning to the left, the opposer was still there. Looking up in surprise, I was astonished to see Gantt smiling roguishly and exclaiming, 'I'm not going to allow you to cut me'."

Along with his love of children can be placed his love of the beautiful. The great out-of-doors appealed to him with all its variety of life, charm, and splendor. In this he was like his hero, "Stonewall" Jackson.

A little to the northwest of Montclair is a stretch of road some five or six miles long, known in Gantt's day as the "Ridge Road," now called the "Alps Road." From this highway the ground slopes downward on either side. Toward the west it stretches across the floor of a wide valley and then rises gradually to the crest of the distant Ramapo Mountains. Time and again, just before the coming of evening, Gantt would drive out to this road, usually accompanied by some friend, timing the journey so that he would reach the highway just before sunset. Then at some convenient point he would park his car facing the west and watch the glory of the setting sun at the distant horizon. He was fond of poetry, could quote it readily, and often his comments would be in the lines of one of his favorite authors, or perhaps in words of his own which were as graceful in phrase and as warm in meaning as any which memory might have brought back to him. After a lapse of years the impressions remaining of his word pictures are as glowing as those of Hoyt:

> Purple, violet, gold, and white,
>   Royal clouds are they;
> Catching the spear-like rays in the west—
> Lining therewith each downy nest,
>   At the close of summer day.
>
> Forming and breaking in the sky
>   I fancy all shapes are there:
> Temple, mountain, monument, spire,
> Ships decked out with sails of fire,
>   And blown by the evening air.

The loveliness of it all had a tremendous, enduring attraction for Gantt. It seemed as if he drew inspiration and strength for his daily work from these moments with the transcendent beauty of the out-of-doors.

Part II

ACTION

FOREWORD

THE narrative record of Gantt's professional life is a story of action. It is dramatic, for it shows the clash of stubborn wills. It is creative, for it deals with inventions and discoveries, some of which will persist as long as industries remain. It is intensely human, for it tells both of successes and failures.

Every activity of life, whether it is an achievement or a disappointment, helps in doing the next task. Each experience sets the work stage for the succeeding act, or measures a step that positions for the next stride forward. The man of real parts ever learns and thus increases his powers and strength. Evolutionary changes may even occur in his personality, and in his technique of doing work, as he solves new problems and overcomes unfamiliar difficulties. Such a development may come from the influence of others, exerted on critical or formative occasions. The mindmarks left by moments like these are often deep and lasting.

Gantt's engineering career shows these features. He grew in ability from the day he left the Institute until the time of his death. His most important technical contributions to industry were made during the last two years of a strenuous life. To his honor he changed his attitudes and methods as he found out others that were improvements. Some of the changes were almost a shift of personality, a recognizable growth and expansion of his mental and emotional powers, which aided in handling both men and affairs.

Part II of this biography is devoted to this engineering and professional work. It tells of numerous positions that he held as foreman, engineer and executive in manufacturing concerns. On one of these jobs he made his contact with the scientific-management movement. During the last eighteen years of his life he carried on a consulting engineering practice, specializing in modern industrial management. He was always busy. Clients often had to wait for his services.

His creative mind produced many inventions. Several dealing with processes and mechanisms were patented. His discoveries in management methods, procedures, and devices, he gave freely to his brother engineers and to industry.

Gantt's great opportunity came when the United States entered the World War. Unreservedly he dedicated himself and his substance to the cause of his beloved country. His contributions to the planning and control of industry during the months of war were so great that many gave to him at that time the high honor of being the *foremost among industrial engineers.*

CHAPTER V

TECHNICAL ENGINEERING

Consulting engineer for economical shop management and for time, cost, and record-keeping." This statement, and a few modifications in phrasing conveying essentially the same thought, were the reference which Gantt made to himself and to his work after he was established as a consultant in modern industrial management. It was in this field that he made his great contribution to his day and generation. Here he accomplished those things for which he will always be remembered and honored, and carried through the thinking which enabled him to leave a vigorous, constructive philosophy of business and industry.

But Gantt did not enter into his management activities at once after completing his education and early training. Like all of the other great leaders in this newer branch of engineering, he did a considerable amount of technical engineering work before devoting all of his energies to the administrative, organizational, and management phases. It is with those developments that this chapter is concerned.

On graduation from Stevens in 1884 he at once returned to Baltimore as draftsman at the plant of Poole and Hunt, iron founders and machinists, then located at Woodberry on the edge of the city. Here he remained for about two years, leaving, as has been sketched in his McDonogh record, to become an instructor in manual training at the preparatory school which meant so much in his life.

During the two years at Poole and Hunt he lived in his mother's home. As the means of transportation were inconvenient, he walked each morning four miles from home to shop, and then made the same journey back at night. His working-hours were such that he had to be in the shop at seven o'clock in the morning, which gives a hint as to his personal habits. He rose early, returned late, and did little but work. It was his habit to go to his room immediately after supper and spend the short evening until bedtime in reading or studying. If, however, he was engaged in a bit of experimental work which grasped his interest, he would often sit up far into the night to carry the research along to a satisfactory stopping-point. Most of his reading at this time was on technical matters. He had not yet awakened to management.

To sketch in brief the contacts covering what has already been characterized as his period of technical engineering: In July, 1887, he completed his second period of teaching at McDonogh School and returned to the shop, or to "metallurgical work at the Midvale Steel Works." This connection continued until 1893, when he became superintendent of the American Steel Car Wheel Company at Garwood, New Jersey. This business relationship was short-lived as the company became financially involved, apparently due to conditions resulting from the industrial panic which began in 1893.

During 1894-95 Gantt was a consulting engineer, with headquarters in Philadelphia, being retained by Benjamin Atha and Illingworth Company, steel manufacturers, of

TECHNICAL ENGINEERING 63

Newark, New Jersey, and the Bodine-Thomas Glass Company of Philadelphia. In 1895 he became superintendent of the Thurlow, Pennsylvania, works of the American Steel Castings Company. This position he held for about a year.

Numerous inventions and several patents are to his credit during the period, 1891-1904. Two of the patents were issued to him jointly with George H. Chase, six were patented jointly with Frederick W. Taylor, one jointly with Carl G. Barth and Frederick W. Taylor, and on eight he was sole patentee.

The Chase-Gantt patents were on a process for casting armor plate and were taken out while Gantt was an employee of the Midvale Steel Works. The patent numbers are 485,784 and 485,785, the date, November 8, 1892. The essential feature of the invention was "the process of producing a steel casting having a hard surface gradually decreasing in hardness from the surface, consisting in providing a dry mold with a lining or facing of a readily fusible alloy, pouring steel into said mold and causing such alloy to be fused and entirely incorporated by alloying with an outer portion of the molten steel."

Several plates were made according to the methods of these patents. Two were subjected to ballistic tests. One of these was made at Midvale, the other was cast by Midvale and forged and finished at Bethlehem. Both were unsuccessful in that, although they broke up the projectiles fired at them, they themselves were broken by the impact. It is highly probable that had these plates been made at a later period, when more was known about the technology of armor manufacture, the results would have been better. No service plates were ever manufactured under the Chase-

Gantt patents, inasmuch as those tested proved inferior to the regular armor then being supplied to the United States Government.

Another invention, Gantt's most important contribution to engineering technology, as subsequent events have shown, was made at Midvale. It was a "mold for steel ingots." The resulting patent, No. 621,646, was dated March 21, 1899. Gantt was sole patentee. The principle involved is basic.

The invention was first tried out and used at Midvale. Later the principle was widely adopted, first in forging-plants, and since 1912 in rolling-mills. The feature of the ingot produced in Gantt's mold is a series of sharp points around the periphery separated by comparatively long depressions of sufficient depth to support the ingot skin in the early stages of solidification. Gantt called it "a contracting ingot mold," in distinction with all others, which expand on receiving molten metal. In the language of the patent specification, "the object of the invention being to prevent the ingots from cracking along the surface, as frequently happens with the molds now commonly in use, the result being that the ingot cannot be properly forged and is therefore useless and must be broken up and remelted."

George A. Dornin estimates that now (1934) certainly 25 per cent, and possibly as much as 50 per cent, of all steel ingot tonnage in the United States is cast in molds designed on the Gantt principle. The amount is increasing year by year. The advantage lies in reducing the cost of conditioning semi-finished steel prior to the final finishing process.

Probably Gantt never received a cent from this invention, or for the use of this patent.

TECHNICAL ENGINEERING 65

Three other patents where Gantt was sole patentee related to improvements in furnaces. These are:

    High-temperature furnaces
        No. 559,940 .... May 12, 1896
    Furnaces for heating and melting iron
        No. 708,705 .... September 9, 1902
    Forge-furnaces
        No. 762,301 .... June 14, 1904

The first two of these grew out of Gantt's work at Midvale; the third from work he did in the forging of metal-cutting tools at Bethlehem.

Around the inventions on heating furnaces Gantt built a short-lived consulting practice. How successful it may have been there is little evidence to show. During this period he turned his attention to the improvement of furnaces for melting glass. It does not appear that the glass-melting furnaces were patented, but Gantt did carry on a business for a while in their design. One of his contracts ended in a lawsuit. The essential characteristic of Gantt's glass-melting furnace development "was the location of both of its regenerators at one end, and its convex or round-nose front which gave ample space for a large number of men to work simultaneously at removing the glass from it." It appears that Gantt turned over certain drawings and specifications for a furnace of this type to the Cox & Sons Company under an agreement whereby it was to pay him certain royalties for each furnace built in accordance with his design. In 1898 he sued for an accounting. The case was decided in his favor and he secured a decree of about $1,500 plus costs.

The Simonds Rolling Machine Company connection was

the occasion for the invention of a ball-grinding machine, patent No. 618,502, dated January 31, 1899. Gantt also developed a ball-hardening machine for the Simonds shop which was a substantial success, but apparently was not patented. The capacity of this machine was about 250,000 one-quarter-inch balls per hour, or, putting it in another way, it would do as much work in an hour as a skilled hardener would do in a day. Another improvement which Gantt made in this plant was a strikingly simple and satisfactory method for testing the hardnesses of steel balls, based on the differences in the resilience of steel of different hardnesses.

Turning now to the six patents taken out jointly with Frederick W. Taylor, they all have to do with determining, measuring, regulating, and controlling temperatures for heating metal-cutting tools. They issued between October 1, 1901, and August 4, 1903. In chronological order they are:

    Metal Bath (Taylor-Gantt)
        No. 683,580 ....October 1, 1901
    Pyrometer (Taylor-Gantt)
        No. 722,770 ....March 17, 1903
    Method of Determining and Regulating the Temperature of Heated Articles or Receptacles (Gantt-Taylor)
        No. 735,361 ....August 4, 1903
    Pyrometer (Taylor-Gantt)
        No. 735,423 ....August 4, 1903
    Method of Determining and Regulating the Temperature of Heated Articles or Receptacles (Gantt-Taylor)
        No. 735,424 ....August 4, 1903
    Pyrometer (Taylor-Gantt)
        No. 735,425 ....August 4, 1903

TECHNICAL ENGINEERING 67

The last joint patent was for a slide rule, patented by Carl G. Barth, Gantt, and Frederick W. Taylor, No. 753,840, dated March 8, 1904. Slide rules such as are covered by this patent have become known as "Barth Slide Rules" and their design and construction have remained largely in Barth's hands. The experimental work, however, which led up to their development, goes back to Midvale days, and first to George M. Sinclair, a graduate of Stevens and a friend of Gantt's. Under Taylor's direction Sinclair worked out simple formulas which expressed with approximate accuracy the effect of each of the numerous variables upon the cutting speed of metal-cutting tools. Of these attempts Taylor wrote (*Transactions A.S.M.E.*, Vol. 28, p. 277):

> The first mathematical solution of the problem was made by Mr. G. M. Sinclair, who devoted, as the writer remembers, a year or more of consecutive work to this end with the help and advice of the writer.
>
> This solution was accomplished by means of overlying curves plotted on ordinary cross-section paper, with which we were able to work out laboriously and exceedingly slowly, for each particular lathe or planer, a set of tables which could be used for most of the conditions met with in ordinary work. This method was, however, so exceedingly slow and laborious as to make it far from generally useful.

When Gantt succeeded Sinclair in 1887 he continued this work. Taylor wrote of this continuing study of the metal-cutting problem as follows (*Transactions A.S.M.E.*, Vol. 28, p. 278):

> After Mr. Sinclair left the problem, Mr. H. L. Gantt devoted a year or more of his time almost exclusively to its solution, and it was during this period that we substituted curves laid out on logarithmic paper for the direct curves laid out on ordinary

cross-section paper. As a result of this work we obtained a logarithmic sheet upon which both diagrams and figures were used to represent the laws, and by means of an elaborate cross slide, upon which further elements of the laws were entered, we were able to make a more rapid and much more direct solution of the problem. This was done, however, by the method of trial and error, but by means of this crude sliding table we were able to make quite rapid approximations to the proper working conditions.

When Taylor once more attacked this problem at Bethlehem, he employed S. L. Griswold Knox, then a member of the faculty of Lehigh University, to handle the mathematical work involved. One result of the joint efforts of Taylor, Gantt, and Knox was to produce "an especially made slide rule accompanied by diagrams, by means of which a still more rapid solution of the problem was obtained."

Somewhat later Carl G. Barth was added to the staff at Bethlehem and placed in charge of all the experimental and mathematical work. This brought him in contact with the work already done by Taylor, Gantt, and Knox, among which was the logarithmic slide rule previously referred to. Barth's description of this instrument is (Barth's Supplement to Frederick W. Taylor's "On the Art of Cutting Metals," Article I, *Industrial Management,* September, 1919):

> ... a combination of a crude or embryonic logarithmic slide rule and a set of tables incorporated on a common body. These tables, in conjunction with the scales of the slide-rule portion of the instrument (which scales were not true logarithmic scales, but merely scales of equidistant graduation of which each mark denoted a 10-per-cent higher value of the respective variables represented than the mark denoting the next lower value of a

TECHNICAL ENGINEERING

variable) embodied all the up-to-then experimentally obtained knowledge of the relations between depth of cut, feed, speed, and life of tool for a certain size and shape of tool . . . together with the several power combinations of the particular lathe for which a particular instrument was specially made up.

Barth's brilliant mathematical mind was intrigued by this problem and, by successive solutions, within six months after he went to Bethlehem he had his final slide rule in practical use in the machine-shop.

To Gantt, however, belongs the honor, along with Knox and Sinclair, of first having tried to solve the complicated problem of cutting feeds and speeds by a graphical method, and his part in the solution of the problem is recognized by the fact that he, with Taylor and Barth, became a joint patentee of the device.

Another group of three patents belong to a later period in Gantt's life. They relate to machinery and apparatus for processing cloth, and were an outgrowth of the work he did for the Sayles Bleacheries. They are described and evaluated in Chapter IX.

In addition to these numerous patents, further evidence of Gantt's inventive ability is revealed by four abandoned applications where he filed as sole inventor, and two where he filed jointly with Taylor. His own were for:

- Methods of Eliminating Oxides from Steel
- Distance Measures
- Apparatus and Solution for Cleaning Articles for Plating Them
- Range-finder

The two jointly with Taylor were for:

- Method of Determining Temperature of Heated Articles
- Flux

Some of Gantt's inventions were patented abroad, with certain reservations on his part as to the wisdom of that course. On one occasion he wrote to his lawyer: "My enthusiasm about foreign patents is getting to be even less than about American patents, which I must confess is oozing very rapidly." On another occasion he expressed disapproval of patent practice, writing (*Work, Wages, and Profits,* p. 53): "One of the foremost American patent lawyers not long ago stated that the tremendous industrial success of the United States had been largely brought about by its beneficent patent laws, and yet the greatest part of the legal talent among the patent lawyers is engaged in evading those very patent laws which are so beneficent to the community.

Gantt's technical engineering ability had been sufficiently indicated. Although facile with a pencil, he was not manually dexterous with tools. He seldom attempted to do anything with his hands. He frequently cut himself while shaving. He was not a good automobile-driver, and it was not always a pleasure to ride with him. He had the unfortunate habit of becoming interested in conversation with some one on the back seat, turning around to make his own comments more effective, and seemingly inviting collisions. In spite of this characteristic he was never in a serious automobile accident.

His record as a shop executive is open to some question. Testimony, which is available to indicate that he was not what would be considered today a good shop foreman, must be judged by the times to which it applies. At about the turn of the century from the 1890's to the 1900's, the prevailing method of handling men in American industry was by drive and force, through arbitrary, autocratic decisions,

TECHNICAL ENGINEERING 71

and by the dominance and will of the executives. It was during this period that Gantt was acting as a shop superintendent.

A personal characteristic enters into this situation. Throughout his entire life Gantt was subject to severe headaches, and when in one of these attacks he was extremely irritable. It is said of him that when he was in one of these moods at Midvale, the men with whom he normally came in contact would tip off one another as to how he was feeling, and if possible keep out of his way until his bad humor had somewhat quieted down. Benjamin A. Franklin, who was Gantt's foundry clerk at this time and thus one of the targets of his irritable outbursts, bears witness to his essential fairness and to his interest in the under dog. He recalls this incident:

> I remember one occasion when he told me he was going to give a sot, who applied for a job, a chance. He said he thought a good hard job would tend to cure him. He put him chipping castings with a big Irishman, the first day, and the big Irishman wore him out, not impossibly on purpose. He then asked for a lighter job and Gantt put him with a red-headed Irishman fitting flasks.
> The Irishman did something Gantt did not like and was taken to task, whereupon he told Gantt where he could go and then quit. Gantt turned to the sot and told him to take hold of the job. Much to Gantt's disgust, he got the same answer from the sot. This set him back a little in his judgment of the advisability of helping the down-and-outer.

In this same period he exhibited many of the qualities which later on became so strong in his management work. One of these was his adaptability, as shown by Samuel H. Libby's description of the first contact with him:

In the spring of 1895 I was located at the Schenectady works of General Electric and connected with the Railway Engineering Department. At that time we were experiencing difficulties with G. E. 800 railway motor castings, the manufacture of which motors was processed in quite large quantities, for the electrification of street railways was increasing by leaps and bounds. The greatest two troubles with which we had to contend were shortage of castings, and castings of uneven thickness, the latter due apparently to a shifting of part of the mold.

Daniel Barton was at that time purchasing agent for the Schenectady works . . . and asked me to go to the American Steel Foundry at Thurlow, Pennsylvania, and do what I could to correct the troubles. Upon arriving at the steel plant I was immediately ushered into the presence of the president of the concern, Mr. Patrick Egan. After I had explained the purpose of my visit to him he sent for his superintendent, who proved to be none other than Mr. Henry L. Gantt, whom I there met for the first time and where was formed a friendship that was to endure until his death.

Mr. Gantt took me out into the shops, where I met the foreman and with him examined the flasks used in making the castings for the G. E. 800 motors, which disclosed two causes for our troubles: the flasks were loose both at the corners and in the fit of pins and lugs; and there were not enough of them for our required output.

Through Mr. Gantt's ready aid, and under my instructions, several flasks were repaired and tightened where necessary, and molds completed and poured. The result was a perfect casting in every case. The trouble was that there was so much looseness, particularly in the guide pins, that the cope was bound to shift in some direction across the drag, and as the core of the drag extended up into the cope, the result was abnormal thickness of one side and a corresponding thinness of the opposite side.

The other trouble seemed not so easily eliminated, for although there were flasks enough for General Electric require-

TECHNICAL ENGINEERING 73

ments, they were being used to produce Westinghouse as well as General Electric castings, and not enough could be made to satisfy either customer.

Again Mr. Gantt came to my rescue, for he and Mr. Watson, their sales manager, went into a huddle with Mr. Egan, with the result that the latter started for Pittsburgh that very night to place orders for a sufficient number of steel flasks to warrant the desired production.

Mr. Gantt, in the meantime, had the old flasks repaired, tightened, and production immediately proceeded with.

Kindly, courteous, Henry L. Gantt was a true gentleman whom it was a rare privilege to count as a friend. He was an inspiration to hundreds of young men who like myself came to know him and who suffered a distinct loss at his passing.

A flash of his leadership is given in a bit of anonymous autobiography related by Gantt in a lecture at Sheffield Scientific School:

> As an illustration of the difference between leading and driving, I may cite an incident that occurred in my presence in a steel foundry. For the benefit of those who may not know, I may say that steel is poured through a nozzle in the bottom of a ladle, and not over the top, as is the case of cast iron. This nozzle is closed with a plug, but for one reason or another this plug sometimes does not close the nozzle entirely after pouring a mold, and the steel leaking out splashes over the ground and the flasks, not only making the neighborhood of the ladle a very hot place, but setting fire to anything combustible within reach.
>
> In order to protect himself from being burnt, should a "bad shut off" occur, the ladleman usually wears thick woolen clothes, including, if possible, an old overcoat.
>
> On the occasion in mind the "shut off," while the ladle was being taken from one mold to the next, was very bad and the splashing and the heat of the molten steel were almost unbearable.

It must be understood that a leaky nozzle is very apt to "freeze" up, not only leaving the molds unpoured, but leaving the steel in the ladle in a large solid mass which it is very difficult to utilize. Moreover, the flasks to be poured are usually needed by the molders the next day, so if they are not poured it is usually impossible to get a full day's work molded the following day.

Notwithstanding these facts, which the ladleman knew perfectly well, he decided that he could not face the heat of the steel from the leaky nozzle, and left his ladle hanging on the crane with the steel running out.

The superintendent, who was standing near, did not say anything, but signaling to the craneman to move to the next mold, went up and, taking the handle of the ladle, began to pour the metal. Before he had finished pouring the first mold the ladleman came up and, taking the handle, poured the remainder of the heat.

The flying sparks had ruined a suit of clothes, but the superintendent had established himself in the estimation of the workmen, and the ladleman, as far as I know, never again forsook his post.

During the period of Gantt's life when much of his energy was being devoted to technical engineering, he joined the American Society of Mechanical Engineers, the year being 1888. Almost at once he began to contribute professional papers and to take an active part in the discussions of the contributions of others. In all he contributed twelve professional papers and sixteen discussions which form a part of the professional transactions of the society. In addition he wrote freely for technical magazines, reviews, and, during the World War, for daily newspapers.

If there are added to these published papers the numerous manuscripts of reports, lectures, and addresses which are available, there is built up a total of upwards of one hundred and fifty of his extant writings. They form a rich heritage from his creative mind. A few of them cover steel castings, furnaces, and fuels. All of the others are on industrial management or related topics, or deal with the prosecution of the World War or with the industrial and economic philosophy which engaged his interest and much of his energy during the closing years of his life.

He felt keenly the responsibility of the engineer toward his professional society. On two occasions when the American Society of Mechanical Engineers went abroad as the guest of foreign societies, he was a member of the official party. These visits were to England at the invitation of the British Institution of Mechanical Engineers, in 1910, and to Germany as guest of Verein Deutscher Ingenieure, in 1913.

From 1909 to 1911 he was Manager of the American Society of Mechanical Engineers, and from 1914 to 1915 Vice-President. He served as a member of the Executive Committee of the society, and a member of the Standing Committee on Meetings and Program, which at that time had the major responsibility for the technical work of the organization.

His active work for the improvement and betterment of society affairs was one of the contributing factors which brought about the appointment of the Committee on Aims and Organization of the American Society of Mechanical Engineers. This group reported in 1920 after Gantt was dead. The recommendations in its report became the basis for the expansion of the society's activities which carried it, before the business depression of 1929, to a position where it had the largest membership and income of any professional engineering society in the United States.

He was also a member of the Society of Naval Architects and Marine Engineers.

As would be expected from a man with Gantt's force, his methods of presentation, both written and spoken, were pointed, vigorous, and concise. On industrial-management topics his writings, more than those of any contemporary engineer, are capable of direct quotation. It is easy to select statements aptly phrased and filled with meaning. They are evidence of the clearness of his thinking and the directness of his expression. These admirable qualities are well shown by a few unrelated selections:

> The truest definition of democracy is equality of opportunity.
>
> A wise policy is of more avail than a large plant, good management than perfect equipment.
>
> ... neither money nor organization will permanently insure success without proper direction.
>
> The authority to issue an order involves the responsibility to see that it is properly executed.
>
> ... the policy of paying satisfactory wages has been more influential in producing low costs than any other item.
>
> Never say a thing can't be done; first thing you know some damn fool will come along and do it.
>
> We should think less about what has been done and who did it, and think more about what to do and how to do it.
>
> ... there is no great ultimate profit in trying to sell a person something out of which he cannot get the value he paid for it.

TECHNICAL ENGINEERING

The average workman is a good citizen just as loyal to his country as the capitalist, and just as proud of its position in the world.

The ideal industrial community would be one in which every member should have his proper daily task and receive a corresponding reward.

... the output of a factory should not bear the total expense of the factory, but only that portion of the expense needed to produce it.

There is another and higher leadership, that of the intellect, by which the methods and thoughts of one man may affect the whole civilized world.

As a matter of fact, it costs almost as much to be idle as it does to work. This is true whether we consider men or machines, or, in other words, labor or capital.

... in the vast majority of people there readily springs up the desire to do something specific if the opportunity offers, and if an adequate reward can be obtained for doing it.

Wealth is convenient, luxury is pleasant; but the nation which does not so develop its industries as to produce men will not for any great length of time hold its place in the world.

CHAPTER VI

SCIENTIFIC MANAGEMENT

As I look back over my own history I can pick out five or six men who have influenced my life more than all others combined; some of these were school-teachers, some college professors, and others were in industry" (*Industrial Leadership,* p. 24).

In these words Gantt gave credit to some of those who had molded and given direction to his life and thinking. In preceding paragraphs certain of these schoolmen and professors have been recognized, and the contributions they made during his boyhood and youth have been appraised. In this narrative events now move along to certain of his associations in industry, and to the individual who, above all others, influenced him amid his shop and factory contacts.

In November, 1887, Gantt went to the Midvale Steel Works to do "metallurgical work," having been recommended by G. M. Sinclair, a classmate at Stevens. His title at first was "Assistant in the Engineering Department." A year later when he applied for membership in the American Society of Mechanical Engineers, he styled himself "Assistant to the Chief Engineer of the Midvale Steel Works" and said he was "engaged in determining the most economical methods of working the machine tools in the machine shop." Within another year he was "Superintend-

SCIENTIFIC MANAGEMENT 79

ent of the Castings Department." This position he held until 1893.

Frederick W. Taylor, the "Father of Scientific Management," was at Midvale during the first three years of Gantt's connection with the works. Then and there was begun a business and professional relationship which was continued in two other companies, for twice when Taylor needed trained, sympathetic assistance he turned to Gantt. More than this, there was a close bond of friendship between them for a number of years until, in fact, a fundamental difference in belief pushed them apart. However, each had a lively regard and genuine respect for the other.

To show the development of scientific management at Midvale when Gantt went there, and the situation into which he entered, it is necessary to sketch in brief outline some of the work of Taylor. Taylor went to Midvale in 1878. Almost at once his abilities were recognized and he began to have executive responsibility. About 1884, the year following his graduation from Stevens Institute of Technology, he became, more or less informally, the chief engineer of the works. In 1887 he was formally elected to that position by the Board of Directors and continued in that responsibility until he left voluntarily in the Autumn of 1890.

It was at Midvale that Taylor laid the foundation for the art and science of scientific management. As early as 1881 time study was established. Then followed functional foremanship. In his book on *Shop Management* (p. 107) Taylor writes: "The writer introduced five of the elements of functional foremanship into the management of the small machine-shop of the Midvale Steel Company of Philadelphia while he was foreman of that shop in 1882-83: (1) the instruction-card clerk, (2) the time clerk, (3) the inspector, (4) the gang boss, and (5) the shop disciplinarian. Each of these functional foremen dealt directly with the workmen instead of giving their orders through the gang boss. The dealings of the instruction-card clerk and time clerk with the workmen were mostly in writing, and the writer himself performed the functions of shop disciplinarian, so that it was not until he introduced the inspector, with orders to go straight to the men instead of the gang boss, that he appreciated the desirability of functional foremanship as a distinct principle in management."

Then came the system of differential rates for the payment of labor. Gantt had a direct contact with this work, as shown by his own comment (*Transactions A.S.M.E.,* Vol. 16, p. 883):

> His [Taylor's] method of fixing rates by elements eliminates, as nearly as possible, all chance of error, and his differential rates go a long way toward harmonizing interests of employer and employee.
>
> It was my good fortune to work for a year as his assistant in this work, and I fully agree with him as to the effect on the men. They improve under it, in both honesty and efficiency, more than I have seen them do elsewhere. Realizing that substantial justice was being done, and that to do their duty was to follow their own interest, it soon became a matter of habit with them.

It was under these conditions and with this pioneer, creative work around him that Gantt had his introduction not only to Taylor, the leader, but also to the principles and practice of scientific management. The power and extent of this directive influence is shown by the fact that a few years later he turned to this kind of work as his own career with results and achievements which made him one of the leaders in industry for all time.

The second contact between Taylor and Gantt came in 1897, seven years after Taylor had left Midvale. Taylor, in the fall of 1893, had undertaken the reorganization of the Simonds Rolling Machine Company. Somewhat later he was retained to improve and standardize the operating departments of the Simonds machine-shop. To give him full opportunity to carry through this work, Alfred Bowditch, president of the Simonds Company wrote the general manager of the works in December, 1896, to inform him that he was relieved of the responsibility for the shop; "Mr. Fred W. Taylor" was to have "entire and absolute authority over the same." This order created great friction, which came to a crisis on June 28, 1897, when the general manager, every foreman, assistant foremen, all the salesmen, and the head men of the office resigned on something like three days' notice. In this situation Taylor turned to his old Midvale associate, Gantt, whom he installed as superintendent.

It was here that William L. Conrad, then a high school boy, entered Gantt's employ. The contact thus formed became a very close association, one that continued to the close of Gantt's life. There were but two breaks, one while Conrad was finishing his education, another when he was in military service during the World War.

The contact of Taylor and Gantt at Simonds was of something less than a year's duration, for in May, 1898, Taylor went to the Bethlehem Steel Company. Taylor's purpose as outlined in a letter written by him to Robert P. Linderman, president of the Bethlehem Iron Company, was (Copley,

Frederick W. Taylor, Father of Scientific Management, Vol. II, p. 11):

> First: To render the management of the shop entirely independent of any one man or any set of men so that the shop will be in a position to run practically as economically if you were to lose your foreman, or if, in fact, a considerable body of your workmen were to leave at any one time.
>
> Second: To introduce such system and discipline into the shop that any policy which may be decided upon by the management can be properly carried out.
>
> Third: To introduce the best kind of Piece Work in place of Day Work and thus stop the loafing which takes place under all Day Work and also to very materially increase the rate of speed and the accuracy of each man in the shop.

On October 23, 1898, Maunsel White joined Taylor, and about November 1 there came the startling discovery of Taylor-White tool steel, or high-speed steel. The introduction into machine-shop practice of tools made from steels of this kind has increased the productivity of the machinery-building industry of America some two to two and a half times.

Immediately after this discovery Taylor instituted a series of experiments to discover the possibilities and limitations of the new steels. With two major problems now confronting him, the organization of the shops, and the research into steel for tools, Taylor turned again to his former co-worker. Gantt went to Bethlehem in March, 1899, "to assist in putting into operation methods for increasing the efficiency of labor."

Gantt's contribution to management methods during his Bethlehem contact is properly a part of the presentation of

"Task and Bonus" in a following chapter. What he did in connection with the tool-steel experiments belongs here. Taylor, in recognizing those who assisted him in his experimental work "On the Art of Cutting Metals," gives this credit to Gantt: "Mr. H. L. Gantt . . . has been interested with us in carrying on these experiments throughout their whole period" (*Transactions A.S.M.E.,* Vol. 28, p. 34).

It does not appear that Gantt, up to the time of the Bethlehem contact, had had much experience in compounding, melting, forging, or rolling high-carbon, alloy tool steel. However, the problems did not disturb him and he attacked them in characteristic fashion. His general method was to set up the desired result as an approximate middle point of the line of investigation, start at both ends, study everything that others had done, check or discard their conclusions by actual trial and elimination, and continue until he had secured the accurate facts of what would work and what would not work.

In this particular situation the problem was, of course, set by Taylor. The solution was largely left to Gantt. Taylor wanted dependable cutting-tools having the property of "red hardness." He wanted tools that would retain temper and a cutting edge at speeds, feeds, and depths of cut far in advance of anything then available. He wanted uniform tools whose performance could be depended upon as one step in setting rates for roughing operations in the big No. 2 machine-shop. The problem might be stated as what to do to increase production of roughing-machine tools by about 300 per cent.

Half a dozen coal-fired holes for melting crucible steel were constructed in the east end of the Open Hearth Building. Bethlehem had no experienced crucible-steel melters or pot-pullers, so Gantt hired two brothers, Frank and John Hoyer, from the Carpenter people in Reading. At the start it was discovered that there was great difficulty in working and forging tools of high-carbon alloy steel. Gantt and White urged further experiments with alloy steels of low carbon content. Taylor would not agree. He claimed that the Taylor-White composition and process patents were limited to high percentages of carbon. Somewhat later foreign "blue-chip" steels came on the market. Most of these were low in carbon but high in vanadium and other purifying ingredients. They proved to be as superior in performance to the Taylor-White steel as the latter had originally been superior to Mushet.

Speaking of this part of Gantt's life, D. C. Fenner, who was associated with him, has this to say: "I have always believed that this was one of the very few instances in which Mr. Gantt was not allowed to follow through his general method of complete and exhaustive experimentation. The result speaks for itself."

Taylor retired from Bethlehem about midyear 1901. Gantt left in September of that same year, and almost at once began his work as a consulting industrial engineer.

Chapter VII

TASK AND BONUS

> They shall work for an age at a sitting
> And never be tired at all.

THESE lines of Kipling's were a favorite quotation of Gantt's, for they presented somewhat his vision of the ideal industrial community in which every member should have his daily task and receive for its accomplishment a corresponding reward. This is what modern industrial management at its best "aims to accomplish, for it aims to assign to each, from the highest to the lowest, a definite task each day, and to secure to every individual such a reward as will make his task not only acceptable but agreeable and pleasant."

In language such as this, Gantt presented his idea of working with an objective as contrasted with working without one. "The idea of setting for each worker a task with a bonus for its accomplishment seems thus to be in accord with human nature, and hence the proper foundation of a system of management." The problem is simply this: To set a proper task and determine what reward shall be paid for its accomplishment.

Gantt's task-and-bonus plan for paying labor was one of the first, if not the first, of many bonus plans for incentive wage payment. Throughout the sweep of the years they have been installed in every variety of manufacturing, and form a method whereby thousands of workers have earned their compensation. They have also been applied to many occupations and in many pursuits outside of those concerned with manufacturing operations. Though no records are available to indicate the number of employees which have been affected, hundreds of thousands have been enabled to earn more under the principle of payment for results, and have increased the amount in their weekly pay-envelopes because of the bonus, rightfully theirs for heightened effort and enlarged production. In writing of this particular development of Gantt's, Dean Dexter S. Kimball says: "To my mind much of Taylor's philosophy would have disappeared had it not been for Gantt. He was the first, I think, to combine Halsey's conciliatory idea of a basic day wage with Taylor's idea of a high reward for a large task. And to make these ideas workable he introduced sound methods of instruction and encouragement. Furthermore, he was a pioneer in extending these methods to manufacturing-plants other than iron-working establishments."

There is no record to indicate that any other industrial management procedure ever was received with so much approval and achieved such an instantaneous success as Gantt's task-and-bonus plan. The story of its introduction is dramatic in the speed of its movement.

The occasion which brought about the recommendation of the task-and-bonus principle, and its first application in the shop, was this. Taylor's work at Bethlehem was moving slowly so far as results were concerned. He had begun the installation of his methods in 1898. As early as 1899 slide

TASK AND BONUS

rules were in use to determine the time for machine work, and instruction cards had been prepared to direct the workmen. However, the monthly output of the shop from March, 1900, to March, 1901, had been but little more than the monthly average for the five preceding years. The record is:

> January, 1896, to March, 1900.... 1,162,418 pounds per month
> March, 1900, to March, 1901...... 1,173,883 pounds per month

Up to this time, that is to March, 1901, the efforts of Taylor and his associates had been directed to studying what could be done. But little work had been attempted to secure the cooperation of the workmen. The men had not helped. Certain conditions were still unfavorable to the introduction of the differential piece-rate system which Taylor regarded as the ideal one for obtaining maximum output. In this situation Gantt believed that it was unwise to expend additional energy and time to obtain the high degree of perfection demanded by Taylor, and suggested that the workmen be offered additional pay in some manner to secure their cooperation.

Accordingly, on March 11, 1901, Gantt's recommendation that a bonus of fifty cents be paid to each workman who did in any day all the work called for on his instruction card, was adopted, and the first bonus rate was set on March 18, 1901. On May 13, 1901, a little short of two months later, R. J. Snyder, assistant superintendent of the machine-shop, reported to E. P. Earle, superintendent of Machine Shop No. 2, on the results secured and the spirit of cooperation engendered:

South Bethlehem, Pa.
May 13, 1901

Bethlehem Steel Co., Department D.M.
Mr. E. P. Earle
Superintendent of Machine Shop No. 2
Dear Sir:

I hand you herewith some notes on the results obtained by the introduction of the "bonus" plan for remunerating labor in No. 2 Machine Shop.

The plan thus far has been applied only to the roughing lathes, and I give below a list of the numbers of the machines with the dates on which they began operating under a bonus or premium:

Lathe No.	76	March 18, 1901
"	" 158	" 19, "
"	" 159	" 19, "
"	" 50-A	" 19, "
"	" 50-B	" 19, "
"	" 145	" 20, "
"	" 146	" 20, "
"	" 160	" 20, "
"	" 208	" 28, "
"	" 207	April 3, 1901
"	" 60-A	" 4, "
"	" 60-B	" 5, "
"	" 90	" 8, "
"	" 30	" 24, "
"	" 55	" 24, "
"	" 72	" 29, "
"	" 73	" 29, "
"	" 4-A	May 6, 1901
"	" 4-A	" 6, "
"	" 34	" 9, "

One of the best results obtained after a short trial has been the moral effect upon the men. They have had it placed in their

TASK AND BONUS

power to earn a very substantial increase in wages by a corresponding increase in their productive capacity, and this has given them the feeling that the Company is quite willing to reward the increased effort. They display a willingness to work right up to their capacity, with the knowledge that they are not given impossibilities to perform. This effect has been brought about by the good use of our excellent slide rules in the hands of a number of the most thoroughly practical men, who, when the results which they demand have been declared impossible to obtain, have repeatedly gone out into the shop and themselves demonstrated that the time was ample, by doing the work well within the limits set. All this has inspired the confidence of the shop hands, and the excellent instruction cards sent out are gradually evolving from laborers a most efficient lot of machine hands.

The percentage of errors in machining has been very materially reduced, which is unquestionably due to the fact that in order to earn his bonus a man must utilize his brains and faculties to the fullest extent, and so has his attention closely fixed on the work before him, as every move must be made to count. He thus has no time for dreaming, which was, no doubt, the cause of many errors.

The condition of the machines is vastly improved. Much care has been taken to point out to the men that the best results can be obtained only by keeping their machines in good running condition, well lubricated and cleaned. They have not been slow to realize this, and cases of journals cutting fast are very rare, while before the introduction of the "bonus" plan this was a very common occurrence. Breakdowns are also of a less frequent occurrence.

The crane service lately has given us little trouble, and lack of crane service was formerly a constant excuse of the bosses and men for not being able to keep machines filled with work. The improvement in this case arose from the rule laid down that no exceptions or allowances would be made for delays due to this cause.

It is only by the introduction of this "bonus" plan that we have had furnished the automatic incentive for men to work up to their capacity, and to obtain from the machines the product which they are capable of turning out. It has lifted the hands of the speed bosses (foremen) and enabled them to act in the capacity for which those positions were created—that of instructors.

These are some of the direct results obtained. Indirectly it has eliminated the constant necessity for driving the men, and has enabled the shop management to divert some of its energy into perfecting the organization, which only will enable us to give a good account of the shop equipment. Much good has also resulted from putting the work through the lots, and in keeping each machine as nearly as possible on the same kind of work.

It is also a pleasure to note in this connection the deep interest taken in this work by the men connected with it, and the fine cooperative spirit which prevails among all hands.

Yours truly,

[signed] R. J. Snyder

The increase in the output of the shop after this plan was installed is shown by the following record, which should be compared with the record previously given for the years from 1898 to 1901:

Average for 1901 . . . . 1,959,130 pounds per month
Average for 1902 2,699,557   "   "   "

The latter figure is nearly two and one-half times the record for the twelve months ending February, 1901, the time when this plan was put into effect. D. C. Fenner credits the major part of this increase to the improvement in management, writing that to the best of his knowledge and belief:

There was no appreciable increase in equipment or productive labor. There was an increase in auxiliary labor, particularly in

TASK AND BONUS

the tool-room and in the shop office, gang foremen, etc. The speed of the main shafting, which was driven by a number of two-cylinder, vertical steam engines, was increased 100 per cent. This was the only physical change.

The differences and points of similarity between the Taylor differential piece-rate plan and the Gantt task-and-bonus plan are put in contrast and comparison by Professor Charles W. Lytle in this way:

> The Taylor Differential Piece-Rate Plan was developed by F. W. Taylor in 1884. He had learned that low labor cost did not assure low total cost. He believed that high wages were compatible with low total cost, in fact, were the means of effecting low total cost provided such high wages were made contingent on high production. This he proposed to accomplish by designating a definite quantity of work for a period of time, that is, a task, and by offering a generous increase in piece rate for all pieces produced in, or in less than, the task time. Such tasks would become standardized and would need to be perfected in advance. Taylor's long studies on the art of cutting metals had laid a foundation of standardized conditions. There remained only the improvement and standardization of the operator's activities. Studies leading to these improved tasks were called time-studies. The essential data for learning these tasks were prepared in the form of instruction cards. The tasks were high relative to what had ever been done before. It was evident that much training and some form of automatic incentive would be necessary to secure their accomplishment.
>
> Taylor wished to offer the equivalent of a cash prize for task accomplishment and he wished to have the prize amply large. For this reason, and because he realized the importance of having only high-class men, he deliberately depressed the piece rate for below-task work. By this means there was room at task for a bonus of 50 per cent without exceeding 125 per cent of normal earnings. From this point, 100 per cent of task and 125 per cent

of base rate, the earning continued as high piece rate. Thus a generous bonus could be earned on all pieces at and above the task rate of production. If an operative should fail to average the task rate of production throughout any day, even by one piece, he must take the lower piece rate. Two operatives could be more than 50 per cent apart in earning, while only one piece apart in performing! Or one man could play the same trick on himself from one day to the next! The skillful, steady operatives liked it, advocated it, and their friends asked to be put on the same basis. But the poorer operatives were dumbfounded. They quit. Worst of all, the trick could catch one at times even if he were good. Machine stoppages, defective material, trouble with a preceding operation, all management deficiencies, could cheat a good man out of his expected high wages.

No plan which emphasizes task point is any simpler to understand or to compute. Like straight piece-work it gives *direct-labor costs*, which are constant, excepting for a step at task. Total *cost per piece* when plotted gives a gradual slope curve characteristic of piece rate, but on account of the change in rate there is a step at task which makes the cost much greater at 100 per cent standard production than at 99 per cent. This, of course, refers to the individual case. When the whole shop is considered, as it must be in judging the merits of any plan, it will be seen that the cost due to those working above 100 per cent standard will be merged with the cost due to those working below 100 per cent. In other words, when the average production of the shop is 99 per cent standard, there will be some individual performances above 100 per cent; and again when average production for the shop is 100 per cent standard, there will also be some of the cost due to those producing below 99 per cent. In changing from a 99 per cent average of one day to 100 per cent average of the succeeding day, there will be no step or pronounced increase in total cost per piece.

Gantt appreciated these advantages of the Taylor Differential Piece Rate Plan, in fact had no intention of substituting any other plan, but he sympathized with the operatives more than

TASK AND BONUS

Taylor did and believed a less rigorous plan could be used while preparing them for higher task habits. He proposed to guarantee a low day wage for all productions less than task. At task he tried a flat bonus of $0.50 but found a step bonus similar to Taylor's more satisfactory. This he varied from 10 per cent to 100 per cent, according to the nature of the work. For machine-tending he used 10 per cent to 15 per cent, for silk processing, which involved eyestrain, he used 30 per cent to 40 per cent; for high-skill work or for skill combined with heavy work he used 60 per cent to 70 per cent. In general he found that bonuses less than 20 per cent were ineffective Above task, that is after the bonus, he first tried an earning curve parallel to the basic piece rate. This anticipated the above-task portion of the Emerson Plan, first used in 1904. Since this curve diverged below the proper piece-rate curve for the given starting-point, Gantt met with some complaint and finally returned to a high piece-rate curve like that of Taylor. Gantt did not, however, stress the piece-rate characteristic of this curve. He always insisted that it was better for the man to think in terms of standard hours accomplished rather than in terms of pieces made. In this instance his ideals outran practice, for the men themselves, it is said, always figured the rate per piece.

Thus, the Gantt Task-and-bonus Plan, as the new plan came to be called, eliminated the drastic or punitive below-task piece rate of Taylor, retained the high-task feature, retained a generous step bonus, and retained a high piece rate for all productions above task. Whatever weakness the plan might have from its guarantee of time wages, was offset by the use of non-financial incentives such as training, production records, and other features of good foreman-operative relations. The term non-financial incentives was not coined until Robert Wolf described his work in 1918, but it is fair to credit Gantt with the first systematic use of such incentives as supplements to financial incentives. The task-and bonus-plan was a success from its conception (1901). By the guarantee, injustice and its component, fear, were largely precluded. Beginners could be put on this plan without embarrassment to either side. At the same time mature operatives could look forward to as high wages as under the Taylor plan.

Direct labor cost per piece combines the figures for time wage up to task and those for piece wage beyond task. Total cost per piece is, therefore, high for low production but quite low for higher productions. Like the Taylor plan, the step at task should be eliminated in considering average total cost per piece.

Taylor at once saw the superiority of the Gantt plan and allowed it to displace his own without regret. Today it is difficult to find any vestiges of the Taylor plan, while the true Gantt plan and innumerable modifications of it may be found in all industries and in all locations.

At the time Gantt advanced the idea for payment of a bonus to the workmen, E. P. Earle suggested that additional compensation should also be paid to the gang boss (the man who supplied the work) or speed boss (foreman) for each day for each of his men who earned his individual bonus. This plan was approved and put into effect. In working it out a further bonus was also paid if all of the men in the department or group made bonus. For instance, a foreman having ten men under him would get 10 cents each, or 90 cents total, if nine of his men made bonus; or 15 cents each, or $1.50 total, if all ten of his men made bonus. The additional 60 cents for bringing the inferior workman up to the standard made the foreman devote his energies to those men who most needed assistance and instruction. Gantt says of this feature (*Work, Wages, and Profits*, p. 115):

> *This is the first recorded attempt to make it to the financial interest of the foreman to teach the individual worker, and the importance of it cannot be over-estimated, for it changes the foreman from a driver of his men to their friend and helper.*

This, in brief, is the history of the development of the

task-and-bonus system which, starting as a temporary substitute for the differential piece-rate plan, supplanted it. The important difference is that the worker who failed to earn the high rate got his day's pay instead of a lower piece rate. Thus, though he was inefficient, he had a chance to earn a living while striving to become efficient.

It was the forerunner of numerous methods of wage payment which use an interval of time as the unit for setting standards and comparing results. This is true for the reason that Gantt insisted on the use of "man hours" or "standard hours" instead of pieces or quantities of work. Thus point systems, and others using time units, have but followed the practice that he established.

The idea and plan of task-and-bonus were born out of Gantt's experience as a teacher, quite as much as from the situation at Bethlehem. He recognized this fact, and was fond of using the child and the lessons of the classroom as analogs in explaining the idea and its applications.

"A good example for our purpose is to study the methods by which a child is taught to perform a simple operation. The invariable method is to explain or show the child as clearly as possible what is wanted and then to set a task for it to accomplish. It may be noted that the accomplishment of the task is rendered much easier for the child and the parent both if a suitable reward is offered for its proper performance. As a matter of fact, setting tasks and rewarding performance is the standard method of teaching and training children.

"The schoolmaster invariably sets tasks, and, while they are not always performed as well as he wishes, he gets far more done than if he had not set them. The college professor finds the task his most effective instrument in getting work out of his students, and, when we in our personal work have something strenuous or disagreeable to accomplish, it is not infrequent that we utilize the same idea to help us, and it does it. The inducement to perform the task is always some benefit or reward. It may not always be as immediate as the lump of sugar the child gets, but the work is still done for some reward, immediate or prospective."

The elements upon which the task-and-bonus plan is founded are (*Work, Wages, and Profits*, pp. 116-117):

> 1. A scientific investigation in detail of each piece of work, and the determination of the best method and the shortest time in which the work can be done.
> 2. A teacher capable of teaching the best method and shortest time.
> 3. Reward for both teacher and pupil when the latter is successful.

Gantt presented the principles of the task-and-bonus plan to the American Society of Mechanical Engineers in a professional paper read at the annual meeting in December, 1901 ("A Bonus System of Rewarding Labor," *Transactions A.S.M.E.*, Vol. 23, pp. 341-372). The discussion was extensive, animated, and complimentary. Charles Day said of it: ". . . the present paper seems to be far ahead of anything yet offered, as it is based on fact. . . ." M. P. Higgins expressed his faith in the plan: "It marks a way for advancing and supplementing all manual skill and technical schooling of mechanics that is scientific and almost ideal."

Taylor in 1903 remarked (*Shop Management*, p. 77): "Task work with a bonus was invented by Mr. H. L. Gantt, while he was assisting the writer in organizing the Bethlehem Steel Company. The possibilities of his system were immediately recognized by all of the leading men engaged on the work, and long before it would have been practicable to use the differential rate, work was started under this plan. It was successful from the start, and steadily grew in volume and in favor, and today is more extensively used than ever before."

Gantt's summation of the results he had obtained presents three points of advantage:

> A very large increase in output, averaging from 200 to 300 per cent;
> A falling off in accidents and breakdowns;
> A quickening of the intelligence of the men.

This professional paper attracted widespread attention. Gantt was invited to prepare articles for several of the more serious magazines, such as *Review of Reviews, World's Work,* and *Engineering Magazine.* He also accepted numerous invitations to speak before various organizations on task and bonus, and continued to do so down to the time of the World War. When in 1913 he wrote his book, *Work, Wages, and Profits,* a considerable part of that disclosure was devoted to the principles and applications of this plan.

In his practice, the installation of these methods became a principal part of his professional work. The soundness of his thinking and the practicality of his methods are shown by the fact that numerous of those early systems are in operation, with very little modification or change, a quarter-century after they were first installed.

As a means of dramatizing to both management and workers the fact as to whether or not bonus was earned or

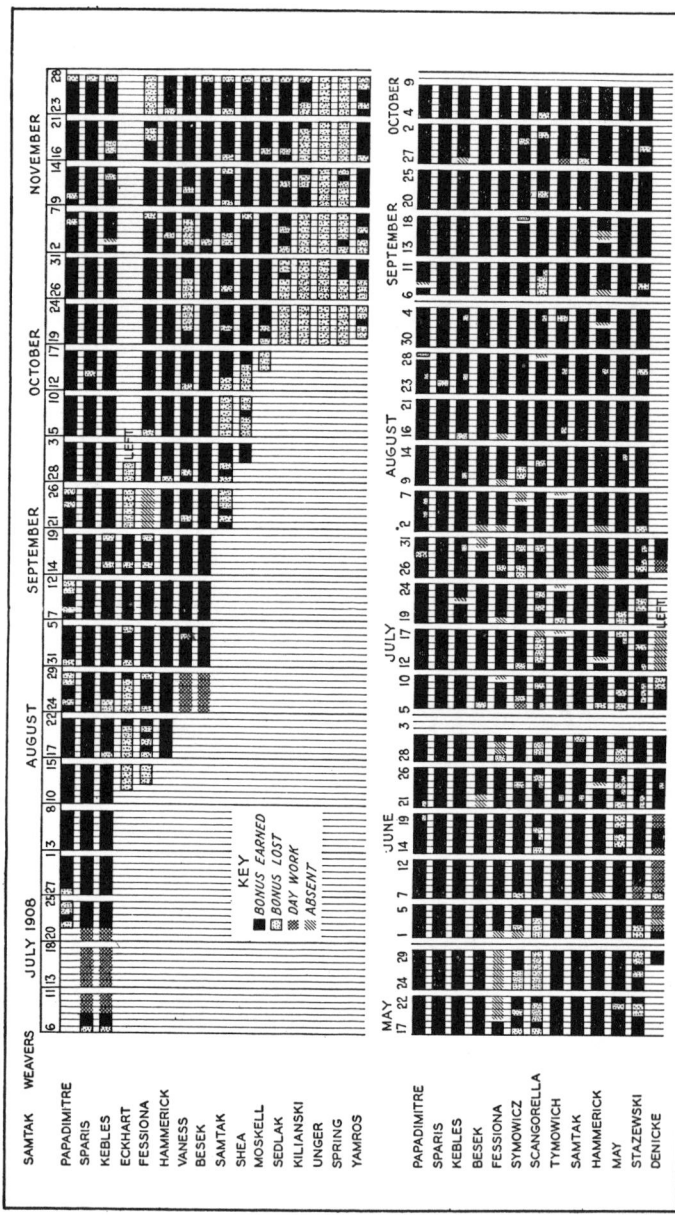

A "Red and Black" Bonus Chart for a Weave-Room

TASK AND BONUS

lost, Gantt developed a bonus chart or "Red-and-Black Chart." A specimen from a weave-room is shown on page 98 (Gantt, *Work, Wages, and Profits*, p. 182).

The story of the application of task and bonus which yields this chart is this (*Work, Wages, and Profits*, pp. 178-182). Three men were chosen at first, those whose names appear at the top of the page. They were Greeks and spoke almost no English. The instructor, Samtak, was a Pole whose English was not very good and who made himself intelligible to the Greeks only through signs.

> The first man, Papadimitri, declined to work under instructions and on task work. He was not discharged, but allowed to work his own way until he should see where his interest lay. We therefore had Samtak give all of his attention to the other two, our observer, who had studied Samtak's work, being constantly on hand keeping a record of the number of picks each loom threw per hour, and removing the obstacles to the men's performing the task. Both men failed to earn a bonus on the first day—this is shown by the red mark—but on the next two days they came so near it that it was allowed them, and they got a black mark.
>
> Our observer, satisfied himself that failure to perform the task was due to the fact that the warps and filling were not coming in a satisfactory manner, and that some of the looms were not just right. He accordingly ceased for a time to urge the men to perform the task, and devoted his attention to getting things in such a condition that these obstacles would be removed. The black cross shows that the men were on day work and were making no special effort to perform a task. At the end of eleven days our observer felt that conditions were all right and he started the men again. Papadimitri by this time had concluded that we were going to "play fair" and wanted to start too. The black lines on the chart show how soon all begin to make their bonus pretty regularly.

We began to study the looms about the first of June and started the first task workers early in July, but it was nearly the middle of August before we were ready to start others on task work. By this time other weavers were willing to try, but it required the attention of both Samtak and our observer to get these men going right. It took the first two of these men about three weeks to become skillful enough to do the task, but the third, fourth, and fifth did it from the start. During September and October several more gradually became proficient.

[The chart] shows the record of Samtak's gang from March 1 to October 9, 1909. This chart distinctly blackens as time progresses. This means more than that the men have acquired the skill to do the work. They have acquired the habit of working steadily and keeping their attention on their work. The red crosses signifying absence are notably lessened. These men have not only improved in skill, but in habits of industry; and the gang boss, Samtak, is not their driver, but their helper and friend. The blackening of the chart signifies not only that more work is done, but that it is done better, for *black* means that both quantity and quality are up to the standard.

The striking results secured by the initial installation of Gantt's task-and-bonus method centered attention on that feature of the development. He felt that insufficient consideration had been given to the method of operating the system by which an exact record of performance was kept. For this reason he prepared another paper under the title, "A Graphical Daily Balance in Manufacture." It was presented eighteen months after the first disclosure at a spring meeting of the American Society of Mechanical Engineers in Saratoga, New York (*Transactions A.S.M.E.,* Vol. 24, pp. 1322-1336). In his introduction to this second paper Gantt wrote:

The paper dealt particularly with the method of setting a

TASK AND BONUS

task and with the reward for its accomplishment. It consisted briefly in setting as a task for a day's work the amount that a good man could reasonably be expected to accomplish, and paying the man a substantial amount in addition to his day's wages if the whole amount was done. If less than that amount was done he simply got his day's wages.

The result of this system, when the task was set in an intelligent manner and accompanied by a suitable compensation, was an efficiency of operation so far beyond that obtained by the ordinary day or piece-work method that it attracted a great deal of attention.

The paper then went on to describe the routine operation of the task-and-bonus method which involved keeping an exact daily record of the work done. The procedure included a combined schedule for work and a balance sheet largely graphical in its presentation. "On it dates are represented by positions, and when work is not done on consecutive days, there are no entries in consecutive positions. This practice enables the foreman or superintendent to see at a glance what work is going along properly." A specimen daily graphical balance sheet is shown on page 103. Gantt's description of it and its schedule taken from this paper of 1903 now follows:

At the heads of the various vertical columns are the names of the pieces to be cast, under each is its pattern number; then, in order, when the pattern is due at the foundry, when it is received, the number wanted per day, and the total number wanted. Below, each column is divided into two columns headed, *daily,* and *total.* These are crossed by horizontal lines representing consecutive working-days, on each of which is entered in the proper column the number of pieces made that day and the total number made to that date. Each column is crossed by two heavy horizontal lines, the upper one opposite the date at which the work should be begun, and the lower one opposite the date at which the work should be completed. These lines are usually red, and have been very appropriately named *danger lines.* The position of the entries with reference to these danger lines and the amounts of those entries show to what extent the schedule is being lived up to. If the schedule is being well followed the entries are always in the neighborhood of the red lines, or above them."

Gantt felt that there would be certain objections to his methods on the score of the difficulty of getting a daily balance. To meet this objection before it might be raised he emphatically declared: *"It is an entirely feasible thing to know exactly all that has been done in a large plant one day before noon of the next, and to get a complete balance of work in order to lay out* THAT AFTERNOON *in a logical manner the work for the next day."*

It appears that Gantt had developed the idea of daily reports of this kind while at Bethlehem. D. C. Fenner, one of his Bethlehem associates, stresses the importance which Gantt, even then, placed upon daily reports of material and labor:

> Our first job, each morning, was to make our labor distribution balance our payroll for the previous day. This, with the report on the distribution of stores had to be in Mr. Gantt's hands by noon. He frequently stated that he placed more reliance upon these daily reports, which recorded the "pulse" of the shop, than he did upon the monthly, quarterly, semiannual, or yearly statements of the accountants.
>
> I do not mean to say that he questioned the accuracy of accountants' statements, but rather that they came too late to be of practical use in planning the daily work and correcting past errors. These daily reports also gave him a pretty fair idea of what the accountant's statements should look like.

FOUNDRY PRODUCTION SHEET A. L. CO. SCHENECTADY WORKS										ORDER NO. 88 8 ENGINES D.L.&W.	
PART	BELL STAND	EXHAUST PIPE	TENDER FRAME CENTER PIN	ENGINE TRUCK SWING BOLSTER	GRATE BAR	GRATE SIDE	GRATE SIDE	ASH PAN END	ASH PAN SIDE	GRATE FRAME SUPPORT	GRATE BAR
PATTERN NO.	17,212	17,939	16,927	16,907	19,458	18,953	18,954	21,343	21,341	18,959	18,961
PATTERN DUE	2-2	2-2	2-2	2-2	2-2	2-2	2-2	2-2	2-2	2-2	2-2
PATTERN REC'D.	1-22	1-22	2-6	2-4	2-9	2-10	2-10	2-10	2-10	2-14	2-14
NO. WANTED PER DAY	1	1	1	1	8	2	2	1	1	1	1
TOTAL NO. WANTED	8	8	8	8	64	16	16	8	8	8	8
NUMBER MOULDED	Daily Total	Daily Total	Daily Total	Daily Total	Daily Total	Daily Total	Daily Total	Daily Total	Daily Total	Daily Total	Daily Total
1903 FEB 2											
3		1 3									
4		2 5									
5											
6											
7	7	1 1 5									
9		2 7	2	2	3	1 P 1	1 P 1	P	P	P	P
10				1	3 3	2 3	2 3	1			
11	4 4			1	4 I 8 10	1 4	1 4	2 3	3 3		
12	2 2 5	1 8	2 4	1 5	7 2 5	2 6	2 6	3	2		
13	1 6		3 7	1 6	7 13	1 7	1 7	2	1 4	4 8	2
14	2 8			1	8 15 17	1 8	1 8	2 4	1 4	4 8	2 4
16			1 8	1 8	1 8 24	1 9	1 9	2 6		4	4 7
17					2 8 30	1 10	1 10	2 8	2 6	1 8	2 5
18					1 8 37	2 12	2 12		2 8	1 1	1 1
19					2 8 46 53	2 14	2 14				
20					2 7	2 16	2 16				
21											
23					1 7 59						1 6
24					5 64						1 7
25											1 8
26											

Begin moulding not later than date opposite upper red line
lower " " " " "
Finish " " " " "

A Daily Graphical Balance Chart

The disclosures in this paper were based on his work with the American Locomotive Company. H. B. Ayers, at that time superintendent of the Manchester Works of the American Locomotive Company, gave this testimony as to the success of the methods: "I have had this system in operation more than nine months and it has worked exceedingly well. We have seen a great increase in our output from its use."

Peck, foreman of No. 1 Machine Shop of the Schenectady works, had this to say by way of commendation: "I honestly believe that Mr. Gantt's system is the simplest, the cheapest, and the best I have ever run across."

Further evidence of the high regard in which his work was held by executives and supervisors of the American Locomotive Company is given in a letter dated July 9, 1903, a few weeks after the presentation of this paper. It is signed by the superintendent and foremen of the Schenectady works of the company, who believe "fully that these methods outlined by you and put into practice here are bound to succeed wherever they are introduced and properly carried out." This letter is reproduced on page 105.

In his closure to this paper on graphical balance Gantt referred again to the difficulty of getting this balance daily, and to the possible objection that the amount of work necessary might make the procedure almost prohibitory. In his characteristic, vigorous manner he met this objection by declaring, "*If it cost fifty times what it does, it would pay.* To know exactly all that was done in a large plant one day before noon of the next, and to get a complete balance of work in order to lay out that afternoon in a logical manner the work for the next day, enables us to *manage a large plant as intelligently as a small one.*"

AMERICAN LOCOMOTIVE COMPANY
SCHENECTADY WORKS
SCHENECTADY, N.Y.

July 9, 1905.

Mr. H. L. Gantt,
    Consulting Engineer,
Dear Sir,

    We, the undersigned, Foreman of the Schenectady Works of the American Locomotive Co., wish at this time to express our appreciation of yourself and your principles of shop management, believing fully that these methods as outlined by you and put into practice here are bound to succeed wherever they are introduced and properly carried out.

    It has been our experience so far that your system points out at once the origination of all delays so that the same may be promptly looked after, and also places the responsibiliy for such delays where it should justly rest, thereby promoting a harmony and good will among employers and employees heretofore unattainable.

    It is our intention to forward the system here to the best advantage, feeling that in so doing we will greatly facilitate and increase the production of the Works.

                          Yours very truly,

LETTER OF APPRECIATION FROM THE FOREMEN OF THE SCHENECTADY WORKS OF THE AMERICAN LOCOMOTIVE COMPANY

This procedure is of importance in any consideration of Gantt's work, for it shows the nature of his thinking which a few years later, at the time of the World War, produced a management mechanism that will always be associated with his name, the "Gantt Chart." The graphical daily balance was one of the forerunners of this more widely used device.

Chapter VIII

SAYLES BLEACHERIES

GANTT's work at the Sayles Bleacheries is outstanding in its record for a number of reasons. The contact was long, extending over nearly four years. Here he fought one of the hardest professional battles of his life, and here he carried over into practice some of his more advanced principles of management. In many ways this experience was crowded to the full with both shadow and sunlight. Here also he trained and developed many of the associates who were most intimately connected with him. Included in this group are: K. E. Adams, David Afleck, T. W. Buscher, W. L. Conrad, George A. Dornin, D. C. Fenner, T. P. Gates, E. A. Lucey, John M. Mullaney and his sons Frank and John, W. E. Pulis, H. P. Reno, R. D. Shaw, C. N. Underwood, C. E. Volkhardt, R. A. Wentworth. In speaking of their work and his own he always used the plural pronoun "we," not the singular pronoun "I." In this way he recognized the contribution of each man to what was done.

If Gantt could have looked forward when the work at Saylesville began he might have seen himself as a traveler standing at the foot of a valley in the depths of winter. The shadowed slopes on either side are covered with snow, the winding brook is ice-coated, the bare trees are swaying in the chilly wind. But in the distance, at the head of the valley, is a hilltop glowing in the warmth of late winter sunshine.

The promise to the traveler is a hard, cold, hazardous journey, then light, warmth, and satisfaction after the difficulties have been overcome.

Among the deeper shadows of the Saylesville experience were a stubborn strike of the employees, the occasion for which was some of Gantt's work; discontinuance of his professional services before the installation of his methods was completed; and the development of his thinking and the extension of his methods along lines which pushed him apart from his former chief, Taylor. On the sunnier side of the experience were the definite improvements and economies which his methods brought about; the building of an organization which had within itself the powers of continuance; and the crystallization of his humanitarian beliefs in regard to the relations between employers and employees into one of the outstanding contributions of his life. This development he gave out in his paper of 1908 before the American Society of Mechanical Engineers on "Training Workmen in Habits of Industry and Cooperation."

Gantt left Bethlehem in September, 1901. In the following January he began his real life's work as a consultant in "modern industrial management." During the eighteen years from that time until his death he served some fifty clients. His great interest in his work led him to take on more accounts than was probably wise even with his robust physical strength. He was governed by the principle that he would take only as much work as he could personally supervise, and it was his practice to visit every job at least once in two weeks, if not oftener. To meet this schedule his days were spent at work, his nights in Pullman berths, his meals were eaten anywhere and everywhere, and finally a none too strong digestion brought his life to an end.

On one occasion he was asked how he secured new clients. He smiled in that quizzical way of his with his lower lip pushed out a little, almost as if he were about to whistle, and said briefly and comfortably, "Oh, they hear about me." After he had written his first book he used it as a means of getting acquainted:

> Before I undertake to do any work for any concern I ask the people employing me or who contemplate employing me, to read this little book, *Work, Wages, and Profits*. I ask those people who have in mind employing me whether they are in accord with the idea expressed in that book of how to handle their workmen, and what share workmen shall have in what is being done. Unless they are willing to subscribe substantially to what I have written in this book, I have always declined to do any work for them.

Gantt was careful in selecting his clients. He would not serve a company with a reputation for oppressing labor. He insisted that the savings resulting from his work should be shared in by the employees. In 1914-15 an Industrial Relations Commission made an investigation into Scientific Management. John P. Frey, a member of the Commission, and later secretary-treasurer of the Metal Trades Department of the American Federation of Labor, recalls several interviews with Gantt. On one of these occasions, after Frey and others had visited the Remington Typewriter plant in Ilion, New York, Gantt stated his policy in regard to accepting clients. The substance of his statement was that "the control over labor given to management by the application of the system which he installed, was so far-reaching as compared with other management controls, that he refused to install his system unless he was convinced that the management was such that no unfair advantage would be taken of the system to oppress labor."

Gantt never charged high fees for himself or his associates. For several years his personal consulting rate was only $35 per day. Herein is striking evidence of his modesty. In fact, that rate was not increased until Charles Day took a contract for him at $100 per day. His method of task and bonus he applied to the compensation of at least some of his associates. He paid them a base salary and a large bonus, in some cases as much as 60 per cent of the income from their work after the base salary had been covered.

Gantt, as the saying goes, "always took his own medicine." That is, the principles which he urged upon his clients he adopted in his own business relations. R. A. Wentworth, an associate, writing from ten years of close contact filled with bright incidents of philosophy, personality, and industrial adventure, has this to say: "Constantly applying to his own affairs those principles which he prescribed for others, he planned far ahead, insisting that visits to clients be arranged weeks or even months in advance. I recall that in June, 1913, in Massachusetts, he made an appointment to meet me at breakfast in a distant city on a certain morning in September, which engagement was faithfully kept without further communication, although his European trip that summer intervened. He made his plans with the knowledge of all the facts, was practically never known to change them, and would not depart from his schedule to suit the convenience of anyone.

Conrad N. Lauer, who became acquainted with him

SAYLES BLEACHERIES

shortly after he started his independent consulting practice, appraises him as "one of the greatest of those contemporary with him in his chosen field of activity." He deserves this high commendation for one reason, because he "built up rather than disturbed the morale of the organizations where he worked."

> His personality [Lauer writes] was a most pleasing one and inspired confidence in all those who knew him. He possessed the rare ability of being able to stand steadfast and firm in the face of opposition, gain his point, and still retain the respect and high regard of those with whom he was compelled many times to differ. There was something unique in this personality of his that made him stand out among his contemporaries as a delightfully human person, one in whom the justice and equity of every situation in which he participated was honestly evaluated. As a result of this procedure, all of those whom he served and those who served with him were impressed with the fact that he was a man of courage in his actions and of absolute fairness in his decisions.
>
> My best opportunity to appraise his work came through association with him on the work which he did for two large and important clients. This work covered many months and resulted in very substantial good to these clients. His ability to analyze these situations and to apply scientific methods to their improvement was at times almost uncanny. His unfailing courtesy and great ability as an industrial analyst made his work all very worth while and of real value to all of those whom he served.

During the first two years after Bethlehem, Gantt served the American Locomotive Company, Robins Conveying Belt Company, Brighton Mills, Williamson Brothers Company, the Portland Company, and the Tabor Manufacturing Company. There seems to be nothing of particular note in this initial period of his professional work. He was installing the Taylor system of management as he had learned it at Midvale, Simonds, and Bethlehem, with the addition of his methods of task and bonus, and a procedure to give a graphical daily balance.

Typical of the work he was doing at this time in his professional career were the methods and procedure that he installed for the Stokes and Smith Company of Philadelphia. This contract was simultaneous with the one at Sayles Bleacheries, but, being smaller, it was finished first.

At Stokes and Smith, Gantt put in a purchase system, stores system, two-bin plan for stowing stores, task-and-bonus plan, organized a Planning Department, symbolized machines, machine parts, operations and cost accounts, and set standard times. The records of this installation are complete. Probably none other of his jobs has such a full set of his reports, with all the accompanying forms and schedules. After more than a quarter-century the purchasing and storeskeeping methods in use are still essentially as he put them in. The task-and-bonus plan, however, has been superseded by a premium method of wage payment.

To turn now to Saylesville. Shortly before the opening of the year 1904, K. F. Wood, superintendent of Sayles Bleacheries, Saylesville, Rhode Island, wrote to Frederick W. Taylor in regard to the possibility of having some efficiency work done at his plant. Taylor recommended Gantt. On January 6, 1904, Taylor and Gantt called on Wood and the contract with Gantt was the result.

The situation at Saylesville as it was at that moment needs a bit of explanation. A few years before Gantt's work commenced, W. F. Sayles, who was the joint owner with his brother, F. C. Sayles, of the Sayles Bleacheries, the Glenlyon Dye Works, and other large holdings, died. His assets and responsibilities were taken over by his son, F. A. Sayles. Shortly thereafter F. C. Sayles retired from the active management of the business and disposed of his holdings to his nephew. This meant that the younger Mr. Sayles, then a man of some thirty-four or thirty-five years of age, was the owner and active executive of a large property in an old industry where there was keen competition.

One of his principal assistants was C. O. Read, who had entered the employ of his father as a clerk and had been advanced until he was general manager of the Sayles Bleacheries. A third executive was K. F. Wood. Wood was a mechanical engineer, a graduate of the Massachusetts Institute of Technology, who had come to Saylesville to assist in developing the mechanical equipment of the mills. He possessed a modern point of view in regard to manufacturing and factory management, showed considerable executive ability, and had been advanced to the responsibility of superintendent. In this capacity he was closely associated with Read.

The internal situation of the mills was typical of the times. It might be described as a group of small kingdoms each ruled over by a foreman who possessed more or less technical knowledge, certain trade secrets, and endeavored to hold his position by keeping this information to himself. Many of these foremen were English; they were ultraconservative, obstinate, and difficult to deal with. Thus the internal affairs of the plant were actually being directed and managed by the foremen and not by the management.

Gantt's coming to Saylesville was strongly resented by this

group of minor executives and supervisors. He had, however, the continued assistance and support of Wood, of Read who relied upon Wood's judgment, and, at first, a substantial amount of support from Mr. Sayles. However, Mr. Sayles was subject to many influences. The idea of an efficiency engineer in a textile plant was novel, and to a number of Mr. Sayles' associates in business unthinkable. Many of these men, who had been friends of his father's, looked upon Gantt's work as a dangerous experiment, and referred to him as "that Pennsylvania blacksmith who thinks he can improve the textile business." The situation was intensified by the fact that among these critics were some of the best customers of the Sayles Bleacheries. Bleaching and finishing cloth are essentially a service business where the skill and artistry of the finisher is relied upon to assist materially in turning out fabrics that are satisfying to the ultimate consumer for their strength, beauty, and usefulness. When the customers of a bleaching and finishing plant become apprehensive over the ability of the concern to turn out goods satisfactory to the trade, it is evident that the entire business relationship is jeopardized.

It is doubtful if Gantt knew all of these facts, or if he did know of them it is likely that he underrated their influence upon his own work. The effect, however, was to hasten the bringing of his work to a close at Saylesville before he had completely introduced his methods. The circumstance of his going made the event an unhappy one for him.

Another factor in this situation was the great expense of Gantt's work. These expenditures were not so much in his personal fees, which were modest considering the work he was doing, but rather for the extensive experimental work which he originated and which cost Saylesville thousands of dollars. When Gantt introduced this work it is doubtful if he had had any previous experience in either the mechanics or the chemistry of converting cotton goods from the gray to the finished state. However, he recognized numerous of the existing inefficiencies and proceeded to search for remedies. To that end he designed and built three machines, the Gantt piling-machine, or "Gantt chute," which is generally used in the bleacheries of the United States today, and has had some adoption abroad, a washing-machine too expensive to operate under conditions of cheap water, and a continuous-boiling machine, or "kier."

Both the piling-machine and the kier were patented by Gantt. Saylesville paid for the necessary experimental work, and, of course, had shop rights under the patents, but there is reason to believe that Mr. Sayles thought the invention should have been the exclusive property of his organization. In fact, he started experimental work to develop a piling-machine which would not infringe the Gantt patent, and after several years of development this result was actually achieved.

Still another factor in the total situation was the death of the superintendent of the dye works at about the time that Gantt came to Saylesville. Mr. Sayles had difficulty, in fact almost met disaster, in finding his successor.

The plain fact is that Gantt had a tremendous number of difficulties, troubles, and responsibilities to deal with, or combat, entirely outside of his work in improving the internal affairs of the company. It is quite evident also that Mr. Sayles had scant opportunity for judging the value of Gantt's work, having to rely principally upon Read and

Wood. There is no evidence that there was a sudden or dramatic break between Mr. Sayles and Gantt. Rather it appears that the dissatisfaction grew slowly until in February, 1908, Gantt was through.

In the atmosphere which has been briefly presented Gantt had trouble to get things done. Many of the foremen left in anger. These were principally old Englishmen who had been valued by Mr. Sayles, Sr., as reliable men. Their replacement was difficult. There were few scientifically trained men who were acquainted with the textile industry. No graduates were available from textile schools, and finishing was looked upon as a matter of personal skill handed down from father to son by tradition. It must be remembered also that Saylesville was Gantt's first textile-finishing job.

The work that Gantt did at Saylesville covered the broad field of shop management. He installed task and bonus, established time study, instituted the setting of rates, developed a planning office, installed a cost system and stores control, developed processes and maintenance methods, created a plant organization which broke up the organization-control by the foremen, and symbolized departments, materials, supplies, operations, and the like. His first work was done in the box-shop. He preferred to make a demonstration of his methods in one self-contained department before going into the larger departments of the bleacheries. In 1913 he wrote of this work (*Work, Wages, and Profits*, p. 212):

> In 1904 we began the reorganization of a packing-box factory, which made five or six hundred cases per day and was run in connection with two large bleacheries of cotton cloth. This factory had been a sore spot, and whenever shipments were delayed, the box-factory came in for its share of the blame. It took nearly

SAYLES BLEACHERIES

a year to get this factory into shape, but for the past nine years it has run so smoothly that the manager of the bleachery has hardly been aware of its existence. In 1910 this factory was running substantially as organized in 1904, with most of the original bonus workers still there.

These box-factory tasks and rates were in use down to 1931, when certain changes of materials made it necessary to modify the standards. Certain other of the rates set by Gantt in 1904 were in use unchanged thirty years later. These facts alone are evidence of the soundness and thoroughness with which Gantt's work was done.

Looking at his Saylesville record more broadly, the organization of the Bleacheries today, 1934, is essentially as he left it in 1908. His methods were extended by the management after Gantt's departure to the Glenlyon Dye Works and are still in use there. Carle M. Bigelow, who went to Saylesville at about the time Gantt's work closed, and thus had an excellent opportunity to know the influence of his work, makes this comment: "The thing that impressed me most was his ability to divorce his mind entirely from the traditions of the industry and, working entirely from the scientific standpoint, set up ideals and ultimately accomplish them, which were almost unbelievable to the practical experienced management."

Gantt made considerable use of his records and results at Saylesville in both his talks and his writings. Two such presentations occur in his book, *Work, Wages, and Profits*. His description of one of these, a red-and-black bonus, gives this information (*Work, Wages, and Profits*, pp. 199-200):

[The chart] represents girls making sheets and pillow-cases. The work of starting the task and bonus was done by a man who had been connected with me, but who was doing this on his own responsibility. I was not personally in touch with this work when it was done. First, note should be made of the fact that the factory was shut down on a number of days—November 28 (evidently Thanksgiving Day), Christmas Day, and all Wednesdays and Saturdays for the next two weeks. It will be seen that the work started off very well, but the rush to get people on bonus on November 30 evidently upset things, for immediately we find a number of workers back on day work. This was probably due to the inability of the task-setter to set tasks on new work fast enough. Note again that just before Christmas week the same condition obtained, and after Christmas there was not enough work to keep the factory running full. However, by the middle of January those that had bonus work were beginning to earn their bonus pretty regularly, and by February 10 the number of workers was just about large enough for each to be supplied with a full amount of work. From that time on the work went smoothly.

In this same book is a specimen of his percentage chart reproduced here on page 119 (*Work, Wages, and Profits,* pp. 207-211):

> [This chart] shows these ratios with regard to some work done several years ago in a bleachery in Rhode Island. Each of the vertical lines represents a different kind of work. It will be noted that the various kinds are represented by the names at the top of these lines. The horizontal black lines marked "100 per cent" represent the amount of work which was done on each of these operations previous to our investigations. The upper lines represent the amount of work now being done, compared with what was previously done. The heavy black line marked "100 per cent" also represents the wages previously paid, and the dotted lines above it represent wages now paid. The 100-per-cent line also represents the previous wage cost. The dotted line below represents the present wage cost. Note that the in-

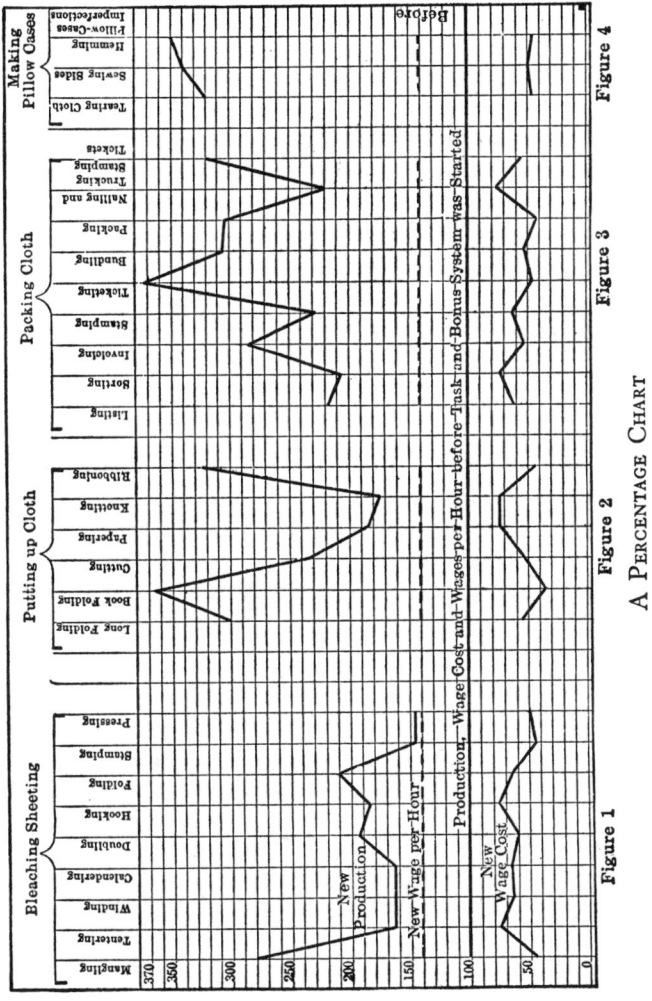

A Percentage Chart

crease in product is about 200 per cent, and the decrease in wage cost is approximately 40 per cent, while the increase in wages is also 40 per cent. Bear in mind that this increase in product is due not solely to the work of the operative, but is much helped by more careful study and cooperation on the part of the management.

It should be noted that this chart represents four different classes of work, all of which illustrate the effect of the task and bonus. In the last three cases the average output is in each instance more than double, and in one, the manufacture of pillow-cases, more than three times as great.

The increase in the case of the pillow-case factory was so great as to make some suspect that the work must have been done very inefficiently before. This was undoubtedly true, but probably not more inefficiently than in many shops run by a foreman who has no special training as an executive and of whom much more is expected than he could possibly do efficiently.

But this is not all; a fortunate set of conditions enabled us to get a measure of the improvement in quality which had been obtained. Soon after the reorganization of the pillow-case factory there was a serious complaint of bad work from one of the largest customers. An investigation proved that the complaint was well founded and the customer was asked to return all the goods.

About fifty cases of goods were returned, and of course the bad work was all blamed on the new system. The inspection of the first few cases proved that the number of imperfections per case varied greatly, and it was decided to keep an exact record of what the imperfections in each case were, and whether the work was done before or after the installation of the task-and-bonus system. The result was as follows: In twenty-eight cases of goods done before task work was started, the average number of imperfections per case was $47\frac{1}{2}$. In two cases done after the task work was started, but before the inspection was going properly, the average number of imperfections per case was 2. In eleven cases done under the task-and-bonus system, after the

GANTT CHUTE OR AUTOMATIC PILING MACHINE

SAYLES BLEACHERIES

inspection was going properly, the number of imperfections per case was less than one.

Representing by unity [on the chart] the number of imperfections per case before the task-and-bonus system was started, the short line represents the number afterward—less than 2 per cent.

This improvement in quality also points to the fact that the pillow-case factory was badly run; the interesting fact is that it was possible to make such a great improvement in a few weeks.

Turning back to the machines whose patents resulted from the Sayles connection: The piling-machine was covered by two patents, "Apparatus for Treating Cloth," No. 911,560 dated February 2, 1909, and "Piling Mechanism for Fabrics," No. 1,074,567, dated September 30, 1913. Application for the first was filed about nine months before Gantt's connection with Saylesville was severed, and the other a calendar month after that event. The purpose of these inventions was to provide a continuous process apparatus for fabrics undergoing bleaching or similar processing. One of these machines built somewhat later in England under the Gantt patent is shown on an accompanying page.

The process of bleaching as Gantt found it he described in a paper presented to the American Society of Mechanical Engineers (*Transactions A.S.M.E.,* Vol. 32, pp. 500-501):

> . . . cotton cloth is generally sewed together piece by piece and handled in the form of a rope which is drawn from one operation to the next by means of rolls. This rope of cloth is subjected to the action of various liquids, being first boiled in an alkali and then washed. After being washed it is usually impregnated with acid (technically "soured"), and allowed to stand in a pile for some minutes to allow the acid to act. The methods of forming this pile and of withdrawing the cloth from the pile, are the operations to which I have special reference.

As the piling operation is repeated after each of several impregnating operations, the successive pilings divide the process into a series of separate and distinct stages with a loss of time between every two. The usual method of piling is as follows:

The cloth is drawn from the souring-machine by an overhead roll, which drops it to the floor beneath. A boy stands on the pile of cloth and so guides it with a stick that it is piled in substantially uniform horizontal layers. When the pile has reached a size determined by the judgment of the bleacher (or the boy), the rope of cloth is broken at a seam and a second pile is formed. When in the judgment of the bleacher first pile has stood long enough, the cloth is withdrawn and pulled through a washing-machine into a bath of chlorine water (technically "chemic"), after which it is again piled in the same manner by a boy with a stick. The judgment of the bleacher as to the time cloth should lie in a pile after impregnation seems to be controlled by his temperament, or by tradition, rather than by knowledge, for we find that hardly any two bleachers have the same opinion as to how long the cloth should be subjected to the action of the acid; and the practice varies from a few minutes to twenty-four hours. As a matter of fact the acid does all its work in ten minutes or less, and no beneficial effect can be discovered by a longer treatment.

Inasmuch as it is necessary to pull the cloth from the top of a pile, the leading portion as it leaves the sour pile has been acted upon by the acid a shorter time than that at the bottom of the pile.

The top of the second pile is attached to the bottom strand of the first pile, and the top of the third pile is attached to the bottom of the second.

As each strand of cloth usually goes through several pilings in the course of being bleached, the action of the bleaching liquors on any portion of the cloth would be alternately long and short, according as that portion of the cloth was at the bottom or the top of a pile. If the rope of cloth was always broken in the same place, the worst that could happen would be an

SAYLES BLEACHERIES

unevenness in the bleach due to the difference in treatment. It frequently happens, however (and this is more often the case than not) that the rope of cloth is not broken in the same place; and when this occurs the various lots of cloth of which the rope is composed, which usually belong to different customers, become almost hopelessly mixed. The expense of straightening out such a mix-up has usually been considered one of the legitimate expenses of bleaching. Add to this the fact that the piling-boy often piles the cloth so carelessly that it tangles as it is pulled off the pile, and not only damages itself, but usually shuts down a portion of the plant for a while.

If we also realize the fact that chlorine (or "chemic") not only forms a most unpleasant atmosphere to work in, but is actually injurious to the lungs, it would seem that some automatic piling-machine which would hold the required amount of cloth and permit the leading end of the pile to be withdrawn would long ago have been devised; inasmuch, however, as this is not a problem requiring a great mechanical skill, but one requiring a somewhat different kind of knowledge, it apparently had never been attacked until the writer came in contact with it.

The piling-machine for treating cloth which he invented and which was the subject of his patents, he also described in the same paper (pages 501-503):

> The machine consists of an inclined chute, with upturned ends, and having a bottom composed of a series of independent rollers, freely revolving. The cloth is dropped into the tall stack and, falling on the rollers, is carried by its own weight to the bottom of the incline. The incline is filled, and as the fabric rises in the receiving stack, the forward end of the pile is forced upward in the other end of the machine, from which it is pulled off at the rate at which it enters the receiving stack.
>
> By making the chute of the proper length, a pile of cloth of any size may be held, and the cloth may be subjected to the action of the impregnating liquid for any desired time, all

portions of the fabric receiving exactly the same treatment. Such action produces uniformity of bleach impossible under the old conditions, and as there is no need for breaking seams, the goods go through the bleach-house in the order they went in, which produces a saving of expense and worry realized only by the man who has operated under both methods. The straightening out of "mix-ups" and the "closing out" of "short lots" are the bane of a finisher's existence, and anything that reduces these troubles does much, not only to smooth the operation of the works, but to assure the customer that he is getting back exactly the goods he sent. Moreover, the dirt and damage caused by the piling-boys are eliminated.

The saving in cost because of the clean, uniform condition of the goods, Gantt claimed was greater than the saving in wages of the bleach-house boys. The relief to the foreman, due to the absence of these boys, permitted him to devote his time to the bleaching process with distinctly beneficial results. In addition to these advantages there was a saving in the floor space necessary for the process.

The development of this piling-machine brought about a more active study of bleaching chemicals and processes, so that today it is common practice to bleach in a continuous operation. This device alone has saved an inestimable sum to the textile industry and has reduced the hazard of non-uniform bleaching.

The boiling-machine was covered in a third patent, No. 1,074,568 dated September 30, 1913. It was applied for some seventeen months after Gantt left Saylesville. It was for a "Processing Apparatus" applying more particularly to the boiling process in bleaching and dyeing. A machine embodying these principles was built and tested at Saylesville while Gantt was there. It was a technical success but an economic failure. It seems unfortunate that Gantt did not have at this time the advice and counsel of textile men who were thoroughly acquainted with the economics of bleaching cloth. Had this counsel been available, and had Gantt accepted it, it is likely that he would not have built either the washing-machine or the boiling-machine. The development of each was a costly experiment unjustified by the final results. Examined through the perspective of time, Gantt was some twenty-five years ahead of his day with these machines, for each has been recently redesigned to meet needs now definitely recognized in the industry.

The piling-machine became a minor source of income to Gantt. He patented the device in England, No. 774-08, and established a British connection which placed a number of the machines in foreign countries. The machine was also patented in France, No. 388,819, and in Germany, No. 224,068.

CHAPTER IX

SAYLES BLEACHERIES (*continued*)

MENTION has been made of the fact that a bad strike occurred when Gantt sought to introduce his methods into one of the principal Sayles Bleacheries. The first attempt at improvement was made in the folding-room. The men folders were folding about 155 pieces of cloth each day for a weekly wage of ten dollars. They objected to the new methods and asked for an increase of 10 per cent in wages. When they could not have their way they struck, and their strike finally carried out the entire plant. The possibilities of improving the folding-shop are shown by the fact that the methods which Gantt finally installed increased the output per day on folding from two to five times.

The feeling against one of Gantt's associates at the time of this strike became so bitter that his opponents sent to New York City for an ex-prize-fighter to come to Saylesville to beat him up. The intended victim was tipped off by a friend as to what was to happen. He had another friend in Erie, Pennsylvania, who was an ex-pugilist. A telegram started this champion on his way to Rhode Island. The New York man arrived first and established his headquarters in a local saloon. Awhile thereafter the Erie fighter appeared, was given a description of his adversary and told where to find him. He went to the saloon, located the New York bruiser, picked a fight with him, beat him up, dragged him out, and put him on a train for home.

At first, the strike of the Bleacheries was a terrific blow to the operation of the plant. Gantt and his associates were compelled to hire promising individuals and by intensive training develop a force of workmen and corps of foremen and overseers. How much this work cost Gantt in personal energy can never be known.

No more complete evidence is needed of the final success of his efforts in developing a new working force, than the fact that the girls of the Bleachery working under the task system organized a "D-4 Bonus Club." At first it was confined to the department which gave it its name. Gradually it spread through the organization, taking in all of those employees who regularly earned their bonus. In this way the workers themselves put a premium on industry and efficiency. The club had its emblem and the girls took pride in presenting one each to Mr. and Mrs. Gantt. The members were so well knitted together that they had their own social affairs, such as dances and picnics. Nothing further need be offered to show how well the employees liked the new conditions against whose installation they had at first walked out on strike.

In sharp contrast to the situation which brought about the Saylesville strike is one of Gantt's experiences that occurred somewhat later in a very large industry. R. S. Gardner narrates the happening, characterizing it as one of Gantt's noble accomplishments:

> At one stage of our work in the Small Motor Divisions, Section 1, of the Westinghouse Electric and Manufacturing Company at East Pittsburgh, we had time-studied approximately 80

per cent of the work. Working-conditions were materially improved and a guarantee had been given to employees that rates would not be cut under the bonus system. In departments other than Section 1 approximately 20,000 employees decided to strike. The workers in Section 1 did not and would not join the strike, which lasted about six weeks. This event was a surprise to the shop executives, strike leaders, and the working-force. It is one of Gantt's outstanding achievements.

From the Saylesville experience there came definitely Gantt's theory and practice of training workmen which is of such importance in his record that the immediately following chapter is devoted to this development. From this time on he insisted upon the function of training and teaching in the development of an industrial organization.

The experiences which led to this change in his methods also created a situation which tended to push him apart from Taylor. Frank B. Copley, in his life of Taylor, contrasts certain differences in attitudes and methods of the work of Taylor and Gantt (*Frederick W. Taylor, Father of Scientific Management*, Vol. II, p. 23):

> The fact would appear to be that such difference of temperament as there was between Gantt and Taylor was creditable to each. If Taylor's slogan was "no responsibility without authority," Gantt told you that as a consulting engineer he wanted neither responsibility nor authority; it was his method to have you come to him that he simply might advise you what to do. It was Taylor's instinct to say, "What ought we to have?" It was Gantt's instinct to say, "What can we do with what we have?" Taylor was thoroughgoing; Gantt did not wish to go any farther than you were willing to have him. Taylor was profound, revolutionary; Gantt adaptable, opportunist. It is true that Taylor felt that Gantt never fully grasped the underlying philosophy of Scientific Management, and that before

SAYLES BLEACHERIES

Taylor's death he and Gantt had largely got out of sympathy with each other as regards various details pertaining to the practice of their profession; but this does not alter the fact that each continued to have for the other a lively regard and respect. It is certain that at Bethlehem, as elsewhere, Gantt's ready ability to make the best of whatever situation arose was of great service in supplementing Taylor's bulldog ability to cling to whatever he undertook until he got there or something broke.

Here Copley dwells upon the difference of temperament between Taylor and Gantt, indicating that it was manifest early in their contacts one with the other. It appears, however, that the relations between Gantt and Taylor were warm and friendly up to about the end of the Saylesville contract. Taylor took Gantt to Saylesville and it is quite likely that Gantt at first kept Taylor informed of the work he was doing. Gradually, however, Gantt realized that the methods which he had been using up to that time were not entirely suited to Saylesville and he found it necessary to modify them. He realized keenly and with deep distress that these modifications were tending to push him away from Taylor. So much was he disturbed by the growing situation that he talked of it very freely to H. P. Reno, who at that time was his principal assistant at Saylesville. Reno recalls conversation after conversation in which Gantt was seemingly with great reluctance accepting the situation that his thinking and beliefs would inevitably push him away from the Father of Scientific Management. The underlying cause was something far more than a difference in temperament. Rather it was a difference in fundamental thinking, in industrial philosophy. In Taylor's attitude was an unwillingness to consider matters except in their technologic aspects. In contrast, Gantt sensed and felt the problems of the general management as well as those of the technological management, and proceeded to take them into account in his practice. Taylor disapproved many of these methods; to him they were compromises rather than true scientific determinations.

One particular point of difference between Taylor's and Gantt's methods is this. Gantt was quite willing to set time allowances on a job after the work had been reasonably well developed and standardized. He recognized the continuous flow of changes which take place in any organization and also that any standard is temporary, never ultimate. He realized that the normal course of human events would bring changes and modifications, and believed that after a reasonable amount of development and standardization work had been done on methods, tools, and operations, a proper moment had arrived to set standard times and put the job on task and bonus.

Gantt deplored deeply Taylor's unwillingness to attach as much importance as he did himself to the painstaking instruction and training of the workers, and to a study of the worker's point of view. He freely admitted that Taylor solved the industrial equation triumphantly according to the factors he recognized and utilized. He had a profound conviction, however, that other factors should be included, and when this was done a somewhat different solution would be reached. However, he could not convince his chief of the soundness of this procedure. Taylor saw no reason for introducing psychological, humanistic, or philosophic factors into a system which he considered as complete and sufficient.

The philosophy which crystallized in this particular bit

SAYLES BLEACHERIES

of Gantt's work became a permanent part of his thinking. He often said, "If you can do a job only 10 per cent well, do it. Then do it again and you will increase your 10 per cent to some higher figure, perhaps 25 per cent. Then keep on, and finally you will gain a degree of efficiency which will be satisfactory to you."

However, Gantt would not yield or compromise on any basic principle. His writings and practice reveal this trait of him. It seems in this respect as if he were continually following the leadership of his beloved headmaster, Colonel Allan.

The occasion of the break between Gantt and Taylor was the former's A.S.M.E. paper of 1908 already referred to, "Training Workmen in Habits of Industry." Gantt submitted the manuscript to Taylor. Taylor returned it with the comment that it ought not to be published. Gantt laid the paper away for a few months to think through the situation. His decision was to submit it to the American Society of Mechanical Engineers for publication. The result is history.

So far as this human situation can be summed up, it appears that Taylor was unyielding in principle and inflexible in practice, while Gantt was uncompromising in principle but adaptable in practice.

Major Fred J. Miller, who knew both Taylor and Gantt intimately, attributes this break in sympathy and friendship of the two leaders to the sharp differences in their origins—protectionist Pennsylvania over against the "Free State of Maryland," in their parentages—ardent abolitionists counter to spirited secessionists, in their early environments—the opulent city in contrast with the plantation and farm, and to critical experiences in industry some of which have been sketched in preceding pages:

> Taylor [Miller is writing] was raised in the Pennsylvania tradition and atmosphere of the great fetish and god, protection to so-called infant industries. Consequently he had little or no realization of the clear economic fact that free exchange of goods among men is nature's great labor-saving device; the first and still the greatest when given free play such as it has had from the first within the great free-trade territory in which we live, and that restrictions of trade and barter by employers or by law at their behest is as utterly indefensible as restriction of production by workers.
>
> Taylor's earlier industrial experience had been such as to cause him to attach too little importance to the *art* as distinguished from the *science* of industrial management; the two being quite different, yet both essential to the best industrial results.
>
> Taylor also failed to realize that this art had been practiced for many generations here and there in many of the workshops of the world, and had been at the same time both ignored and scoffed at in many other workshops. Where it flourished one found not scientific management as we know it today, but did find satisfied, enthusiastic, interested, and devoted workers giving the best service they knew how to give, voluntarily, without driving, but with true leadership nevertheless. Where this art was absent there was constant grumbling and dissatisfaction on both sides, the workers complaining of unfair practices, brutal driving, and disregard of what are generally recognized as fundamental human rights. There was large labor turnover, usually disregarded because believed to be inevitable or of little importance, both erroneous as we now know. Unions were fought in every possible way, employers often doing collectively the very things they so bitterly condemn the unions for doing or attempting to do.
>
> In contrast, Gantt clearly recognized basic economic facts

which, whether we like it or dislike it, do in the end govern practically all of the doings of men in their commercial and industrial activities. He clearly understood both the art and science of industrial management. Of course, he did not consider it to be his task to educate his clients in the fundamentals of economic science and did not attempt it. He was nevertheless guided by such fundamental principles and thereby avoided many errors such, for instance, as bitterly assailing workers and especially the unions, for restricting production, and making public addresses that seemed to be based upon the assumption that there was little or no restriction by others.

To show Gantt's convictions on these matters, Miller quotes two excerpts from *Work, Wages, and Profits,* pp. 20, 27-28:

> In considering the subject of management we must recognize the fact that in this country, so long as a man conforms to the laws of the State he has a right to govern his own conduct, and to act in such a manner as his interests seem to dictate. Granting this, it follows that any scheme of management to be permanently successful must be beneficial alike to employer and employee, and neither labor unions that regard their interests as essentially antagonistic to those of employers, nor employers' associations whose only effort is to oppose force with force, can ever effect a permanent solution of the problem of the proper relations between employers and employees.
>
> Labor unions demanding all they can get, and employers' associations organized simply to oppose the demands of the unions, can never evolve a satisfactory system of management; for although each, in its way, may be (and undoubtedly is) often beneficial to its members, both are formed with the idea of using force only, which can never be a substitute for knowledge.

Gantt's generous nature did not permit him in any way to undervalue or to disparage Taylor and his greatness. A short time after the latter's death, in a lecture at the Sheffield Scientific School, Yale University, he spoke this generous tribute (*Industrial Leadership,* pp. 27-29):

> There is another and higher leadership, that of the intellect by which the methods and thoughts of one man may affect the whole civilized world. Industrial leaders who have most prominently attracted our attention in the past are those who have, by their inventions or their direction of activities, accumulated large fortunes; but none of these are as great as the man who by the force of his intellect leads people throughout the civilized world to benefit themselves and others. Such a man was the late Frederick Winslow Taylor, who, in his determination to eliminate error and to base our industrial relations on fact, set an example which will have an effect all over the world.
>
> His great contribution to the world's work was to substitute knowledge of human activities for opinion as a basis of action.
>
> His insistence that all industrial questions could be best answered by a scientific investigation was at first scoffed at by many of our industrial leaders, and it was nearly twenty years before he got much support. Now, however, at the end of thirty-five years his persistence is bearing fruit so rapidly that the whole industrial world is undergoing a revolution due to his ideas.
>
> His death cut short the activities of a man who had the welfare of his fellow man at heart, and who spent much of his life in trying to establish a basis on which the relations between employer and employee could be made mutually satisfactory.
>
> When he began his work, almost all such relations were established by opinions. Today there are few industries in which fact has not supplanted many opinions.
>
> He had the feeling that waste was a crime, and that efficiency in work was a duty not only to ourselves and to our employers, but to the community at large.
>
> His name will live as that of a man who could rise above

individual cases, and grasp general laws that would make for the happiness and prosperity of all.

In an address delivered at a memorial meeting to Taylor, Gantt made reference to the challenge which Taylor felt in an unsolved problem, the thoroughness with which he sought for its solution, and to Taylor's belief that a strenuous life was the life worth while, in that "it added to the usefulness and happiness of men." He also quoted lines as expressive of one of Taylor's beliefs, which, with equal force, might be applied to Gantt himself:

> That when the day is over and your work is all well done,
> That when the campaign's ended, that when the battle's won,
> Then friendship keen, and memory of many happy days
> Bring the glorious satisfaction that a life of action pays.

Gantt's dynamic personality and dramatic method of expression are still remembered at Saylesville. Two such recollections are these: One of the sore spots in the operation of the bleacheries was the maintenance of the machinery. This unsolvable problem was turned over to Gantt. He picked out two machinists who were ordered to patrol the plant, make minor repairs, and issue repair orders for more extensive repairs when they were necessary. One of these men was too efficient. He turned in a large number of repair orders. Gantt tried to restrain him and get better results, with little effect. He could not fire the man directly, for Gantt himself was not a shop employee. One day this incident occurred. The over-zealous machinist had turned in several particularly indefensible orders. Gantt cornered him, told him of his shortcomings, and said: "If I had written out orders like these, I would resign. Yes," striking one fist into the other palm, "I would resign at once." The machinist quit. Satisfactory managerial control of the maintenance problem was finally secured by offering a bonus to the patrolling repair men for those days when they did not send in repair orders. That is, the success of their work was shown by their ability to make the repairs themselves and not ask for assistance from the machine-shop.

On another occasion, a young Englishman had been hired in the planning office. After he had been at work a week, he asked Gantt for a raise in wages. Gantt jumped on a table or desk, attracted the attention of the fifty or sixty persons in the room, and said: "Here's a man," pointing him out, "who has worked for us one week and wants a raise in pay. What shall we do with him? Has he had time to learn his job and our methods in only one week?" The applicant for a raise faded away.

No more fitting closure can be made to this narrative of Gantt's work at Saylesville than testimony as to the persistence of the fundamentals he taught and the permanence of the methods he installed so many years ago. H. P. Reno, who was his particular assistant for a considerable part of the time, and who is now, 1934, vice-president of the Sayles Bleacheries, writes:

> The changes which in the last twenty years or more have transformed almost everything that has to do with our industrial life, have altered many things in our plant also. Yet if Mr. Gantt were again to walk briskly through the familiar door where we had grown accustomed to look for him, and were to traverse the route through the factory that he took thirty years ago, he would find himself among familiar things—surrounded by evidences of the methods he had installed. He would find time study, bonus, stints, schedules, detailed instruction

sheets, symbols, cost system, and here and there a workman who thirty years ago with misgivings started to work "on bonus" and who all these years has enjoyed working "on bonus." He would even find some time allowances established by himself so long ago still in effect, except as they have been adjusted to take care of minor changes in material or methods. And he and we would find ourselves carrying on discussions on problems of management, using the same language—and in agreement on fundamentals.

Chapter X

TRAINING WORKMEN

APPRECIATION and approval, the like of which are seldom accorded to an engineering idea, were freely given in the discussion of "Training Workmen in Habits of Industry and Cooperation," a paper which Gantt presented to the American Society of Mechanical Engineers in December, 1908. He must have been both gratified and encouraged at this appraisal of his work by his brother engineers, but in his closure he confined himself to emphasis upon the place and responsibility of the engineer in his relations to the mechanic, on the one hand, and to the capitalist, on the other.

To show the nature of the reception given to this paper, William Kent valued it so highly as to place it in the forefront of all those which had ever been presented to the Society: "The hopeful thing about this paper, which I regard as the most important that has ever appeared in the Transactions of the Society, is that it is in harmony with humanitarian ideas."

Charles Piez also emphasized the humanitarianism of the plan: "What appeals to me most in Mr. Gantt's presentation is its distinctly human tone; the spirit of helpfulness toward the worker which it evinces. He recognizes that people as a rule are willing to work at any 'reasonable speed and in any reasonable manner if sufficient induce-

TRAINING WORKMEN

ment is offered for so doing, and if they are so trained as to be able to earn the reward,' and he finds in the application of his system that 'an *instructor*, a *task*, and a *bonus*,' prove most useful."

Conrad Lauer referred to the results which are obtained if a spirit of cooperation is engendered in the workmen: "It is the spirit of helpfulness which runs all through Mr. Gantt's paper that especially prompts the writer to pronounce it well worthy of serious consideration."

H. V. R. Scheel commented on the increase in efficiency and improvement in economies which such a plan would bring about: "No one can doubt that such a system of management must result in the greater efficiency of workmen, individually and collectively, and in the greater cooperation of workmen, foremen and managers, with the attendant economies."

Dr. Alex C. Humphreys was impressed with the ethical influence of the methods: "It is encouraging to see the stress laid by the author upon the ethical influence of the methods he describes; and I venture to believe that, if this system were generally introduced throughout the United States, the resulting moral uplift would attract more attention than the increase in dividend-earning capacity."

In the opening paragraph of the paper Gantt referred to the widespread interest in the training of the workmen which had been marked for several years in American industry, and ventured this criticism: "The one point in which these methods as a class seem to be lacking is that they do not lay enough stress on the fact that workmen must have industry as well as knowledge and skill." *Habits of industry,* he maintained, are far more valuable than any kind of knowledge and skill. "Without industry, knowledge and skill are of little value, and sometimes a great detriment."

He stated the purpose of the kind of training he advocated in these words: "If workmen are systematically trained in habits of industry, it has been found possible not only to train many of them to be efficient in whatever capacity they are needed, but to develop an effective system of cooperation between workmen and foremen." He then reminded the reader that he had outlined the basic principles of training in a previous paper on "Task and Bonus," and in this presentation was concerned mainly with the methods of application which he had found advantageous in his own experience. "The general policy of the past has been to drive, but the era of force must give way to that of knowledge, and the policy of the future will be to teach and to lead, to the advantage of all concerned. The vision of workmen in general eager to cooperate in carrying out the results of scientific investigations must be dismissed as a dream of the millennium, but results so far accomplished indicate that nothing will do more to bring about that millennium than training workmen in habits of industry and cooperation."

The body of the paper was devoted to the practical application of the principles, through indicating what must be done and how it is best accomplished. The closing paragraph deals with the subject of moral training: "The fact that under this system, everybody, high and low, is forced by his co-workers to do his duty, for some one else always suffers when he fails, acts as a strong moral tonic to the community, and many whose ideas of truth and honesty

TRAINING WORKMEN

are vague find habits of truth and honesty forced upon them. This is the case with those in high authority as well as those in humble positions, and the man highest in authority finds that he also must conform to laws, if he wishes the proper cooperation of those under him."

Any adequate consideration of this paper must fit it into its times. Following the disastrous business panic of 1893-99, handicraft industry had disappeared in the United States. About 1900 American industry was completely organized on the factory system. The attitude of executives toward the workmen was, in general, arbitrary and autocratic. The management planned everything and ordered the workers around. They were expected to obey without objection, and to work without assistance in doing their tasks. A man was supposed to know his trade or job and to be able to perform the work assigned to him. Against this harsh, stubborn attitude of management, employers, and owners Gantt boldly declared for a policy of teaching and helping instead of ordering and driving, for treatment which would win and hold the employees' cooperation instead of arousing their antagonism and ill-will.

It is difficult to realize how revolutionary this doctrine was in 1908. At that time American industry had not passed through the evolutionary change brought by the World War, nor had it suffered from the depression that soon followed. It was striving with sheer force to build the greatest institution of industry that the world had ever known. In this situation Gantt's voice was like that of a prophet, alone and unheeded except in his immediate circle. A few of his associates, friends, and fellow engineers realized the bigness of his concept and the constructive character of his doctrine. Otherwhere it is doubtful if this great contribution that Gantt made to American industry was appreciated or appraised at its true value.

Nearly a decade passed before the training of working-people became a generally accepted practice. The occasion for the change was the needs of industry during the World War. Several million young men were taken out of industrial pursuits. A great demand came for materials of all kinds. Industry absorbed everyone who could help, young and old, rich and poor, men and women. Many of these persons were strangers to any kind of industrial work. They not only lacked a knowledge of manufacturing, they had no machine sense. To meet this situation there grew up the practice of training working-people the like of which had never existed before in American, or any other, industry. Then it was that Gantt's teaching of a decade before began to bear fruit. It is, to be sure, quite impossible to connect directly his paper of 1908 and the tremendous growth in employee-training of 1917 and 1918. Yet Gantt had clearly defined the principles and developed the practice which, through direct adoption, intelligent adaptation, or rediscovery, became a major part of the industrial management mechanism of the years of the World War.

It is possible, in its final effect upon American industrial practice, that this paper of Gantt's had a more far-reaching effect than any other of his contributions. Such situations are difficult to evaluate and appraise. But this is true, that the training of workmen as it was developed within a decade after the presentation of this paper has become an integral part of the management practice in American industry.

TRAINING WORKMEN

In the chapter dealing with Gantt's Saylesville experience, it was pointed out that this paper was the direct outgrowth of his experience in rebuilding the working-force and corps of supervisors of the Bleacheries after a disastrous strike. That period of stress brought something else into his practice, a radical change in his technique of installing management methods. There is no better evidence of this change than to place in contrast his purpose in the work he did at Brighton Mills before the Saylesville experience and his objective after that connection was ended.

Alfred F. Ernst, who worked under him in 1906-07, and again in 1908 and following years, gives this impression of the change:

> My impression of the first years of Gantt's work at the Brighton Mills is that his main purpose was the increase of production both per worker and per machine in order to cut overhead and expand the margin between expense and income.
>
> My later contact with him in connection with the studies relating to idle machinery gave me the impression that while he was aiming to secure a greater return on the investment, he was bringing emphasis upon the responsibility of the sales department in keeping the plant supplied with orders. He emphasized the advantages of securing low cost by larger quantity of production from the investment and particularly, I remember, emphasized that concerns might expect to remain alive and to compete by a reduction in price based upon a lower cost of operation of a fully utilized plant. It is also my impression that while Gantt was keeping in mind the necessity of a return upon the investment, he was also thinking of the benefit such a reduction in price would have socially, and, as I recall some conversation, he thought of the advantage of this not only to the worker, but also the person of so-called fixed income.

W. O. Jelleme also had an unusual opportunity to observe this change in Gantt's work. He describes and explains it as follows:

> In September, 1906, I started work for Brighton Mills as timekeeper. Mr. Gantt had started his work a few months before. The first efforts were to obtain cost and timekeeping records. Previous cost records had been run on the basis of average yarn number and average poundage cost. The effort then under way involved the introduction of time cards on a daily basis, and one of my duties was the changing over of the carding and spinning departments from weekly time cards, punched daily on a clock, to the card-a-day job time card favored by Gantt. The new time-card system was part of the effort to classify all work and get job costs. It included an extensive set of mnemonic symbols for all expense classifications, and also a plan for cost accumulation for all products. All employees on direct labor were required to change time cards, which were then time stamped whenever changing from one product to another. In the cloth-inspection department this meant changing cards on the average at least once an hour. All cards were then figured separately for payroll. For cost accumulations another sorting and totaling of the cards gave the costs.
>
> The purpose of that first year was entirely getting at information, or, to use Gantt's oft-repeated expression, "Let's find out." In that year's experience I saw a mill partly operating on the basis of every foreman a law to himself, with hiring and firing, payroll, and stores under his own control, and partly on the basis of an organized, planned control of information. Control of operation had not yet been considered. That year's work was for Gantt only an opening wedge to get information with which to begin his real work.
>
> I left Brighton in 1907. Soon after, all consulting work by Gantt was stopped. I returned to the mill in the summer of 1909. By that time Gantt had again been retained and work

TRAINING WORKMEN 145

was in full swing. During that summer I worked as investigator on various projects of getting the facts, and began to have a contact, at first only indirect, with Gantt and his work. By this time the first time studies had been made and the first task and bonus set in the weave-room. The higher pay and lower cost resulting from task-and-bonus operation brought demands for more from both labor and management. It also, of course, put the spotlight on the poor balance of production between departments and the need for planning and routing. In returning to the work Gantt had insisted on going all the way. The first work done had showed the direct influence of Taylor and of Shop Management. The second phase, which included the introduction of planning and routing, began to show departures from the strict job set up of the Taylor-system type. The job costs set up at first had been tried out and found of little meaning for practical purposes. Time cards were still being changed, but it was realized something else had to be developed. By the summer of 1910 a thorough cost analysis based on factors was under way.

The growth of Gantt's work and influence through this early period, as contrasted with the later period when he returned to Brighton, were a measure of the growth of the man and of his thinking. A basic foundation had to be laid in accumulated facts. Even allowing for that, it was apparent that the early work was based on tangibles, time study, setting of task and bonus rates, and cost accumulation. Management principles and the organization of a business as a whole apparently had little to do with it. The responsibility of management as a serious consideration was beginning to be apparent. The first approach was entirely by shop mechanism to organize the workmen. In the later period it was by organization of the management. The first assumed that more work, lower costs, and higher wages could be simultaneously obtained.

The effort was very direct and centered on direct labor and routing materials. From task and bonus in the weave-room it spread to various pay incentives for all parts of the plant. Various experiments were made with the speed of machines to find the best speeds for the purpose. Based on comparatively short tests as they were, many of these findings had to be reversed later. Machines would not stand up under the speeds, or quality would not hold up. For many years the Brighton spinning was notoriously poor running, largely because of crowding the preparatory machinery. A mania for speed and for crowding production had resulted from the savings obtained from Gantt's early experiments, and had been carried too far.

Due to the concentration in this period on the elements of shop management, the tendency in Brighton was frequently to over-emphasize the relative importance of these items and to under-emphasize the importance of the intangibles of general management and financial management. As the base was built, however, this was generally overcome, and a better conception of management problems came about. In fact, the beginnings of this came from Gantt's insistence on finding *why* every time a weaver did not make his bonus. That focused attention on the shortcomings of the management in poor supply of materials or poor condition of machines, and the like.

In the period from 1906 to 1912 Gantt saw the problem as one of improving the efficiency of labor by job analysis and planning the flow of materials. The market was always two jumps ahead. By 1915 or 1916, when he returned for further consulting work, he was thinking of getting the most out of the investment. Prices were rising, and the company that had no idleness of machinery could show the best costs. Because of rising prices, plant idleness assumed greater importance and the utilization of fixed assets meant greater production per dollar of investment.

In 1915, in a paper, "Modern Methods of Training Workmen," Gantt gave his beliefs in regard to the part that the state should play in training the worker in manual dexterity. From this paper a few paragraphs are excerpted:

TRAINING WORKMEN

The widespread adoption of the public-school system has committed our country to the responsibility of training our youth intellectually, and the time seems rapidly approaching when the state will assume the responsibility for training the youth in manual dexterity. There is no question that this is the logical outcome of our industrial conditions, and one of the problems which faces us is just how far it should go in special training. In other words, if the state accepts the responsibility for industrial training, how far shall it accept the responsibility for vocational training?

Just as some knowledge of engineering and of industrial processes is rapidly becoming one of the essentials of a liberal education, so also is an elementary knowledge of the use of the ordinary tools of our common industries becoming an essential part of any education.

It is my feeling that when our school system has given this generous training, it has assumed all the responsibility for the training of workmen that can be legitimately put upon it. Any additional training must have special reference to a particular industry, and is generally termed vocational training. Such training it is the function of the industries themselves to give; but, in order that a workman may develop himself to the best advantage, vocational training should always be preceded by industrial training, which gives him the ability to learn more than one trade with surprising ability, and thus develops in him a spirit of independence and self-reliance, the value of which it is hard to over-estimate.

In another paper, written within a few months of his death, Gantt turned to a consideration of workmen's training as a part of the foreman's job:

> What we are trying to do here is to teach the men, who do not know, how to produce more. It is the foreman's job to remove the obstacles in their path and to help them. A man may have many different reasons for not doing so much as we think he should, such as no material, cranes busy, trouble

with tools, etc., most of which he cannot correct. Foremen in the past have often felt that if the man could not do a job, the thing to do was to fire him and get one who could. This method is ineffective for not only is it often an injustice to the man, but we do not get the results we are after. Our plan is to see if we cannot teach the man how to do better and more work. This will create a better feeling in the shop, make better mechanics, make them more satisfied with their jobs, and make a better living for them.

If a man does not do a piece fast enough, the foreman should remove the obstacles, the majority of which are beyond the man's control but entirely within the control of the foreman or superintendent, and then show the man how to do the job. We do not approve of foremen "cussing" their men, but we do approve of their showing the men how the work is to be done. Ninety per cent of the men will respond to good treatment by their foremen.

The general principle is: Help the man who is most in trouble; then the man who is next in trouble will come and ask you to help him. We shall thereby get a spirit of coöperation. If we do that the shop can pay wages enough so that everybody can get a good living and be continuously successful.

Gantt was fond of "showing up" the top management of an industrial organization and placing responsibility upon its members for the work they ought to do. Training workers, he declared, is as much a function and responsibility of management as providing satisfactory machinery and suitable materials.

Chapter XI

REMINGTON TYPEWRITER

"THE largest, most pleasant, and most successful job I have ever made." Such was Gantt's judgment of his professional work for the Remington Typewriter Company. He was thinking of the changes, improvements, and betterments that had been installed with his counsel and advice, and the ensuing advantageous operating results. Of equal significance are the changes which took place in Gantt himself during this consulting connection. They were sufficiently revolutionary to constitute almost a change in his personality as exhibited in his professional relationships.

The Remington job is noteworthy for still another reason. It was the occasion for cementing one of the great friendships of his life, and thereby brought a directive influence to bear upon him and his work. Major Fred J. Miller, the friend who exerted this influence, was the general factory-manager for the Remington Typewriter plants. He selected Gantt as the adviser for his company when he proceeded to improve the industrial operation of the properties under his charge. He brought to Gantt's support a matured, ripened, both intensive and extensive, knowledge of American industry, and a firm but kindly feeling toward everyone in his organization. These qualities in Miller had a mellowing effect upon Gantt's attitude in business relations, and thus brought about another improvement in his technique of installing his methods.

This change in Gantt is remarkable for the fact that he was forty-nine years old when he began to work for Remington. That he did change, and change so decidedly, is evidence of the fundamental soundness of the man, of the elasticity of his mind, of his ability to efface himself, control his own emotions, and unite his personality with the group in which he was working.

Mention has been made of Major Miller's extensive acquaintanceship with American industry. Included in his experience were fifteen years of shopwork and twenty years in technical journalism, during ten of which he was editor-in-chief of the *American Machinist*. He had traveled widely both in the United States and abroad and had visited hundreds of machinery-building plants. During these contacts he had talked with all kinds of factory executives and had observed that the tolerant, patient, fair-minded manager had little trouble with his employees. When Miller gave up journalistic work he returned to industry as the general manager of all the manufacturing operations of the Remington Typewriter Company. Its five factories in two different states employed some 5,000 workers. The products were typewriters, their accessories and supplies.

Miller had met Gantt frequently at the Engineers' Club in New York City, and in this way had gained a knowledge of his work and somewhat of his way of thinking. When the time came to select a consultant, Miller at once turned to him as the best man of his acquaintance for the post. It is interesting to note that this selection was made without consulting Taylor, although at that time Miller was much better acquainted with Taylor than he was with Gantt.

Gantt's engagement at Remington began in 1910 and continued until 1917, or for about eight years. The impression he made on a close observer at this time in his life, Miller describes in this way:

> A distinct characteristic of him was his unassuming manner. He never "put on any airs" and seemed never to try to impress others with himself; but only by what he advocated and believed in. He was entirely sincere and absolutely honest with himself and with others. He paid little attention to his personal appearance, had good clothes, but took little care of them. I think few ever talked with him, even for a few minutes, without realizing that he was not only very well educated, but was a thinker and a very unusual man.
>
> He did his best work, however, when working by himself and writing out his thoughts. He was not a very good extemporaneous speaker and seldom did himself justice in such speaking. His writings are far superior to what he said orally.
>
> His work, when deliberately done, was always very well done, but he was at times impulsive and hasty in his judgment of others, and sometimes he would appear to explode and "bawl out" men with whom he came in contact in the course of his work, men who had perhaps failed to fully understand some point in his previous instructions and asked him a perfectly proper question about it.
>
> In some cases it seemed impossible for him to believe that a man to whom he had explained a thing should not fully comprehend it thereafter, and a question showing that he had not been fully understood, or that the man had simply forgotten a point, was liable to upset him. Sometimes he thought the question was asked in order to puzzle or annoy him; and in that way he alienated some men who could easily have been made friends and cooperators.
>
> But he was fundamentally just and fair-minded, and when he

gave himself time enough would generally come to right conclusions regarding the character and abilities of those with whom he came in contact, being, moreover, always quite frank and willing to say that he had been mistaken.

Two of the fundamentals in Miller's policies adopted when he undertook the managership of the Remington plants were: First, that he would discharge no one if that act could possibly be avoided, and that he would allow no one to persuade him to take such a course; second, that he would secure the support not only of the officers of his own company, but also of his subordinates and assistants before taking any important step.

Gantt tended to clash with these fundamentals in Miller's policy as soon as the Remington work commenced. It must be remembered that he began this work only a short time after his unhappy release from Saylesville. The memory of those years of bitter struggle were still fresh and keen in his mind. Vigorous methods, an expression of his own intense will, which had carried that job to a substantial success, were still a part of him. They were the cause of incidents at Remington like this:

> One day he stood at one end of the drafting-room in one of the factories and assailed the superintendent, who was standing at the other end of the room, a performance which no doubt promptly reached the ears of nearly everybody in the factory, and set them to wondering "whom they were working for."

Another, as related by an eye-witness, is:

> Gantt was a firebrand. I well recollect his going into the office of the punch-press department with Mr. Gallivan, who was supposed to be in charge of the production procedure, and in Mr. Gallivan's presence upbraiding the production clerk in the

punch-press office in no uncertain terms for the way the production scheme was handled. The clerk, a faithful citizen by the name of Fred Mowatt, was badly hurt by this, as he had placed Mr. Gantt on a pedestal and was prepared to worship him. Seemingly Mr. Gantt used that procedure in bawling out the underlings to carry home to the superiors the fact that the job was not being carried on properly.

News of these and similar rough acts came to the knowledge of Miller. With tact, patience, and at the same time firmness, he suggested to Gantt that it would be better to leave all matters of discipline to him. He took the ground with Gantt that he had a well-known status in the organization and was more likely to be with it a longer time than Gantt himself. Gantt saw the force of this reasoning and soundness of the position taken, and agreed to Miller's request.

Clarifying incidents of this kind, continuing conversations between Miller and Gantt, and the development of the work as it progressed brought about the change in Gantt which has already been characterized. So far as Gantt's contact with others was concerned, the change was felt principally in his technique of installing his methods. Thus Miller's influence was shown in two directions—in Gantt's attitude toward those around him, and in his approach to the problem of putting in his methods and procedures.

To avoid any misunderstanding of Gantt's attitude toward work and management, emphasis must be given to the fact that he never "bawled out" a workman. He would talk with him with the greatest freedom, but on his level, as man to man. His attacks, when things did not go well, were always made upon a supervisor or executive. Even then he usually found a way to apologize and smooth things over at the end of the incident. Often, after such a happening, he would turn to an associate with this half-apology, "Well, I made my point, anyway."

To explain what occurred a little more fully. Gantt's ideas and methods were fundamentally sound and he believed in them whole-heartedly. When he went to Remington he attempted at first to get them accepted by forceful presentation, as he had done before. He would bully foremen in the shop, shout at them until he could be heard from one end of the plant to another, and often "call down" not only his clients' men, but his own staff members, without caring who was present. Gradually, all this changed. He came to the conclusion that he could get farther in having his ideas accepted by appealing to the interest, even the self-interest, of men with whom he came in contact, than by the methods he had used in the past. He learned by experience the tactical advantage of a quiet, private talk with a man whom he wished to influence. In spite of the habits of years, he reversed his manner of dealing with people. Throughout the period of transition it was interesting for his associates to watch the inward struggle between his autocratic tendencies and his new point of view.

Gantt had been the first of the pioneers in management to attempt to humanize the science, that is, to fit methods to the average workman or shop foreman, and to take into consideration their states of mind. He always insisted that the worker was the variable, and that all else must be adapted to him. In this respect he had most of his colleagues in the profession against him, but in Miller he found one who not only understood his point of view, but who could help him carry his purpose farther than he had thought possible. Miller's calm and unemotional judgment, and his insistence on taking into consideration all opposing points of view before arriving at a decision, had a profound influence on Gantt's approach to new problems.

Thus the Remington experience softened Gantt's impulsiveness, developed in him a stronger emotional control, gave him a greater kindliness in dealing with subordinates and factory executives, and the opportunity to install his methods to the full. Once his control over himself had been established, Gantt was an easy individual to work with, for he was fair-minded and at heart well disposed. Although he took great pleasure and pride in the success of his Remington work and did his part with comparative ease, when the results were achieved, with the utmost freedom he gave a large share of the credit for what had been accomplished to the Remington factory organization.

Gantt was consistent, but at the same time he was not afraid to change. It might be more correct to say that he was continually driving forward into new fields. Step by step he would organize the knowledge he had gained, standardize the method of handling that kind of a situation, and then would insist that his staff must follow that standard method without deviation, unless there was something very unusual about the conditions in which it was to be applied. With one matter safely behind him he would move on to something new.

During the latter part of his Remington connection Gantt developed the practice of holding an informal gathering in his office in New York City on Friday afternoon of each week. Almost anyone interested could come. In particular he sought the presence of key men in the organizations of his clients. In these meetings counsel and advice were freely given and problems were brought forward for discussion and solution. Throughout these sessions Gantt seemed to be once more a bit of the school-teacher, very much the industrial leader, and something of the philosopher. He was carefully thinking things through, but was prepared for instant action when the occasion demanded.

Wallace Clark, one of his associates, gives us an insight into Gantt's treatment of younger men during these years, and the inspiration and assistance which he afforded them:

> I used to go frequently to his office in the Singer Building for guidance and encouragement. The fact that he was no longer paid for advice made no difference, and he would spend long hours discussing a certain situation and guiding me toward the solution of my problem. One of the things that struck me particularly at that time was the way in which Gantt classified each situation which was described to him, and the accuracy with which he pointed out or suggested the causes. The problems which I brought to him were intensely personal to me and were confused with individual prejudices. He quickly placed the problem in its proper class and showed me that the clash was not due to personal peculiarities, but to differences in points of view or aims. He helped me to see that my task was to present any proposed course of action in such a way that it would not clash with the self-interest of the men whose support I needed. It was in these conversations that I began to get the first glimmerings of the professional point of view in management, to look at my problems objectively. It was hard to detach myself from them, for in some cases the opposition took the form of personal attacks which threatened the loss of my position and my income.
>
> Gantt was always very kind to any young man who came to him for advice or help and, in return, would ask the young

man to listen to the newest idea he was turning over in his mind. He did not work at his ideas in the quiet of a laboratory. They were forged in the heat of action in the shops, shipyards, or wherever he was working, and then they were hammered out and tempered in conversation with anyone who would lend a sympathetic ear. He would explain the idea to one person and get his reaction and perhaps turn the phrase a different way. The next day he would try it on some one else, each time digging up something to add to it from his own mind or that of his listener. That was one of the secrets of his clear, forceful speeches and writings. There was no surface brilliance, but a clear, ringing truth in every sentence.

Gantt was an ideal teacher and was able to bring out the best in his men. He was never completely satisfied with anything we did, always making us feel that we could have done better. On the other hand, none of his criticisms was personal, but always grew out of the work which needed to be done. He would devote hour after hour to helping us understand the problem and its solution. If we did not finish at the office he would ask us to go out to Montclair for the evening or for over Sunday, and we would continue the discussion there until everyone but Gantt was exhausted.

His interest in young men extended beyond their work and business relations. F. D. Manning, one of his younger associates, recalls this bit of typical kindness:

> I remember that while I was on the J. H. Williams job I was going to be married. I wrote to him asking how much time he felt I could be away, but at the same time I realized that, as my services were charged only for the number of days that I worked, the time that I was away took a definite amount of income out of his pocket. His reply to me was entirely characteristic of the man. He inclosed a sizable check as a gift. He stated that I could take as many days on my honeymoon as I wanted. He closed his letter by saying, "Your expenses will bring you back."

Chapter XII

REMINGTON TYPEWRITER (*continued*)

THE first work undertaken by Gantt at Remington was to regulate the operating procedures of the factories. Then followed a study of the flow of materials, from their purchase and receipt, through all of the manufacturing operations on to the finished typewriter. Miller did not delegate any administrative responsibilities to Gantt or to any of his associates. Thus the problems that Gantt was called upon to deal with were those of plans, methods, procedures, and performances. He had a hearty dislike for the word "system" in such a relationship. He recognized that in the minds of too many persons it was a catchword, an abracadabra to work wonders without due effort, and a cloak to cover a multitude of sins.

The complexity of the problems he attacked is indicated by the fact that a standard typewriter, as then made, contained about 2,500 parts. The several Remington factories were producing 481 distinct typewriters, aside from special machines, with 1,113 different authorized and listed keyboards. Machines were made for all the written languages of the world and for many dialects. In some of the manufacturing departments tons of steel were used every day, as, for instance, in the section where the key levers were produced. Other parts weighed no more than a tenth of an ounce, and if only one of these were used on a machine,

REMINGTON TYPEWRITER 159

two or three days' output would make a year's supply. Between these extremes were thousands of other parts requiring a varied number of operations, and with manufacturing procedures carrying them through various departments. However, if a single part were missing, no typewriter could be assembled or shipped.

After securing control of the operating procedures and the flow of material through the works, Gantt next devoted his attention to the "tight spots" in the factories. These were the places where it seemed difficult to get the work through to keep up with the rest of the flow of production. In practice they are often indicated by the necessity for working overtime on certain jobs. By studying and improving these jobs first he was able to make immediate additions to the output of the factories. This result helped in keeping the peace with those who had an aversion to what was called "red tape," and with those others who were afraid of what they called "unproductive labor." At the same time studies were made of machines and groups of machines by men with stop watches instructed to note every stoppage of work, for any cause whatsoever, its cause and its duration. Gantt then undertook to eliminate the unnecessary stoppages and to shorten as much as possible the remaining ones. In doing this he found out many interesting facts not previously known, or, if known to the workmen, not known by those who could do anything about them. These discoveries were by no means confined to shortcomings of the workmen; they were just as apt to be shortcomings of the management. Gantt taught the importance of having complete control of all operating procedures before beginning time study and bonus work, because the whole organization must effectively and unfailingly cooperate for the accomplishment of tasks.

Two years were spent in this preliminary work of getting control of things. This was by no means lost time. The cost accounts showed that it paid and paid well.

The major results achieved after nearly seven years' work were these:

1. The invested capital had been reduced relative to the volume of product. This was brought about largely by a cutting down of the quantity of raw and finished material in storage and of work-in-process. Other factors were, the sale of one plant, amalgamation of its products with that of another, and a reduction in the number of machine tools and amount of manufacturing equipment required.
2. The factory production had been increased in the proportion of 100 to 164. While this record was being made only a few special manufacturing machines were added. The number of working-hours per week had been reduced from 59 to 50; the number of employees per unit of product had materially lessened. The actual working-time had had still another reduction by the introduction of a ten-minute rest period into each half-day.
3. The wages paid to employees had increased in the proportion of 100 to 123. At the same time there had been an average saving of 18 per cent in one factory on the labor cost of the jobs studied and placed on task-and-bonus.
4. The average labor cost had been reduced in the proportion of 100 to 80. In the largest factory, time study and bonus at the close of 1916 had been applied to 3,607 different jobs. Production on these jobs had increased an average of 65.5 per cent. The average wages of the workers on these jobs had increased 24 per cent.
5. A steady improvement in the quality of the finished machines had taken place during the installation period,

REMINGTON TYPEWRITER

6. All factory employees and executives had been carefully trained for their duties.
7. Both executives and employees had the satisfaction of knowing that they were working under a well-thought-out plan in which each man's duties were defined, and one which provided for the advancement of those who made good.
8. From top to bottom there had been cultivated a spirit of coöperation and good will among all the executives and employees.
9. The factory executives had been relieved from the supervision of routine affairs, and made free to deal with the exceptional matters which needed careful study before decisions were made.

The relations of important operating factors, comparing 1912 and 1916 performances, were:

Year	Machines produced	Employees	Payroll in dollars	Average wage rate per hour in cents	Working-hours per week
1912	86,000	2,700	$1,765,000	.24	59
1916	92,000	2,100	1,760,000	.34	50

This record is all the more impressive for at Remington Gantt had to meet the highest standards amid which he had ever worked. The plants were building an interchangeable product, produced by a good organization, excellent tools, well-developed methods, and a high-grade personnel.

One of the earliest steps in the installation of Gantt's methods was to divide or classify the work as an aid in securing control:

> The first class included all the shop work that is routine in its nature; that is, all the work that can be done from previous experience. This work was directed and controlled from the "man-

ufacturing office," which had the duty of determining "when" and "how many."

The second class of work included that which is new; that is, things which are entirely outside of the routine and concerning which the shop has had no previous experience or exact knowledge. The making of all kinds of shop investigations, engineering work of the plant, taking time studies and setting tasks were all put in this second class. Control of these items was centralized in the "engineering office," which had the duty of answering all questions of "when" and "how."

This subdivision of the work was basic in all the factories where Gantt's methods of shop management were installed. The reason for it was to separate that which is done over and over again and becomes routine, from that which is done for the first time and must be carefully studied step by step. Each class of work required a different type of man; the first, one who is content to do the same thing time and time again; the second, one who prefers to do a thing but once and is always interested in that which is new.

Gantt's fondness for terse statements is shown by the way in which he set up the prime functions for the main operating procedures.

Prime Function of Storekeeping:
1. To have on hand, when wanted, all material needed.
2. To see that all material not needed is disposed of.
3. To safeguard all materials not issued to some special order.
4. To keep in the manufacturing office an exact balance of material on hand and on order.

Prime Function of Timekeeping:
1. To know what workmen worked.
2. To know when each workman worked.

REMINGTON TYPEWRITER 163

3. To know what each workman did.
4. To know what each workman was paid for what he did.

Conditions for Task and Bonus:
1. The workman must have a constant supply of work ahead.
2. Proper instructions must be furnished as to what is to be done and how it is to be done.
3. An accurate record must be kept of the amount of work done and the time it has taken to do it.

In the planning department at Remington Gantt worked out still another step in his system of graphics. Miller gives us a description of this phase of the procedure of control:

> In each factory there was established a central planning department, which controls the beginning and every subsequent movement of a lot of parts through our factories. It has charts showing the capacity of every machine in the factory—how much it can do in a unit of time. It has charts showing to what date or time the capacity of every machine has been taken by work already assigned to it, and when planning the progress of a lot of work through our factories, the jobs are distributed among the available machines as shown by those charts, to the best advantage. One chart shows when each operation upon a lot of parts should be completed, and the daily reports show how nearly this time has been met. Thus any serious failure to maintain the schedule is at once detected, in time to do something about it and to avoid serious delay. Also all interference of one job with another is done away with, and the planning department knows positively that when it arranges a schedule for a lot of parts to go through our factories, those parts will go through in accordance with that schedule, unless some unforeseeable accident occurs. And when such an accident does happen, then is the time for appropriate executive action until the schedule is restored. Thus all foremen are relieved of the work of planning, which, usually, they are ill fitted to do. They are not in possession of the facts for broad planning; they haven't the breadth of view, and know

Cost and Production Charts Developed at Remington

Chart 1: Carriage Head Screw Bushing

NAME	Carriage Head Screw Bushing						ORDER NO. PW 32700-23	
DATE ORDERED	9/16/15	PIECES ORDERED	2 P P	3 P P	10 M	PIECES RECEIVED 8900	PIECES DEFECTIVE	
DEPT.		1 P S						
OPERATION		Auto-Screw Machine	Burr	Slot	Case Harden			
	MATERIAL	DOLLARS CENTS	DOLLARS CENTS	DOLLARS CENTS	DOLLARS CENTS	DOLLARS CENTS	DOLLARS CENTS	DOLLARS CENTS
DATE								
10/18	178	66						
		88						
		80		14	26			
		33		74	82			
		87						
		91						
		09						

SUMMARY OF COST

TOTAL MATERIAL	1	78
TOTAL LABOR	7	66
EXPENSE @ %		80
TOTAL COST	9	42
		105

Chart 2: Back Spacer Key Stem

NAME	Back Spacer Key Stem						ORDER NO. PW 74900-25
DATE ORDERED	7/3/15	PIECES ORDERED	15,000				
DEPT.		1 P P	2 P P	3 P P	4 P P	5 P P	6 P F	7 P P
OPERATION		Blank	1st Form	2nd Form	Pierce	Co-sink	Harden	Assemble
DATE	DAILY TOTAL	DAILY TOTAL	DAILY TOTAL	DAILY TOTAL	DAILY TOTAL	DAILY TOTAL	DAILY TOTAL	DAILY TOTAL
7 21	4500						
	4900 9400						
	3100 12500						
26		645					
		5200 5845					
		4885 10730					
		1700 12430					
31			3870				
			3500 7370	5085			
8 2			4200 11570	3020 8105	2250		
				3300 11405	9100 11350		
7						11300	
9						11300	
14							3280
							3350 76 30
							3520 11150
							100 11250

little of the relation of their departments to the other departments of the work.

A specimen of one of these operation, or production, charts made out for a "Back-spacer Key Stem" is reproduced on page 164. It is a graphic production record showing several operations. Each day's production and the accumulated total are given, thus establishing a direct connection between time and output, and tracing the sequence of operations.

The upper portion of the chart gives the cost of producing a part. The piece for which this chart was prepared is a "Carriage Head Screw Bushing." The costs of three operations, in addition to the material cost, are shown. This type of chart in its graphic plotting is the same as for the production chart already described.

On this type of chart, for certain of the operations where the parts are handled in bulk, the number of parts is not given, but a straight, horizontal line properly located as to date indicates the completion of the work.

Chapter XIII

CHENEY BROTHERS

"I HAD very close relations with Mr. Gantt and believe he contributed a very important amount to the industrial development of the country." Such is the testimony and appreciation of Horace B. Cheney, vice-president of Cheney Brothers, written some fifteen years after Gantt's contact with his plant had ended. At this factory two important developments were worked out. The one, the application of Gantt's methods of analyzing idleness, the other, the plotting of another of Gantt's charts.

Charles Cheney, treasurer of the company, wrote to Gantt on January 25, 1912, addressing him in care of the Engineers' Club in New York City, saying he was looking for a better factory system, and asked for a conference. At that time Gantt was so busy that, although he was quite ready to undertake the contract, he was unable to begin work until early November. This professional contact was long, continuing without interruption until his energies were absorbed in work for the United States Government during the World War. Thus it was one of the longer periods of professional services that he rendered. The work was mutually satisfactory, has persisted, and is still the basis of the management methods in Cheney Brothers mills.

The four principal features which he introduced are:

CHENEY BROTHERS

Task and bonus, which, in this installation, was the most fundamental of all

Time study

Instruction and training of the individual worker

Job analysis, with the establishment of reasonable allowances for delays and difficulties in getting work done

It was Gantt's habit to invite his friends to view his jobs and see the work he was doing while it was in progress. Professor Joseph W. Roe accepted one of these invitations and visited the Cheney mills at South Manchester, Connecticut. During the visit, Roe asked Gantt, among other questions, where he made his start in such a situation—that is, when he came into a plant with a free hand to install his methods, what was the first move he made. Gantt's response was this: "It is right here in the room where we are standing." The place was a very large weaving-room with many scores of looms. Previous surveys had convinced him that here was the heart of a textile plant. Furthermore, it was apparent that there was a low percentage of use factor—that is, the ratio of actual machine hours to the possible number was small. Continuing, Gantt said: "My first move was to put a series of observers into this room and when any machine stopped to determine immediately the cause and run it down." When making these investigations he assured both the workers and the foremen that he was not going to disturb them at all, but that he simply wanted to find out the causes for the delays and to remedy them. The major cause, it might develop, was away off in another part of the plant, in the dye house. An order might be all ready for setting up, with the material there, patterns, men, and looms all ready, and all of the colors called for on hand but one, yet both machine and goods would be held up until that one color could be supplied. Remedying this lack and other faults raised the percentage of use of the equipment in this department, which was the heaviest earner in the plant, from some such figure as 58 per cent to somewhere around 90 per cent.

During that same visit Gantt in his terse, direct way summed up to Roe the essentials of good management: "Two-thirds of all the difference possible between an obsolete, inefficient management and the best possible one lies not in time-study work, wage-payment systems, complicated functions of control, etc., but *in having all the material when you want it, where you want it, and in the condition you want it.*" Then, with a twinkle, "This last sounds simple, but to accomplish it requires two-thirds of the work anyway."

Gantt's ability to write direct, concise statements which, on the one hand, could not be misinterpreted, and on the other urged action, is shown by the reports which he made on the Cheney job. Two of these, taken together, are excellent examples of his ability in this respect. One, dated May 29, 1913, is a short report on the ribbon mill. The other, dated September 18, 1913, is a report on the various mills.

RIBBON MILL REPORT

South Manchester, Connecticut
May 29, 1913

Tubular Ties:

    Get balance on remaining commissions as fast as possible.
    Enclose storeroom for ties.
    Start schedule and order of work for inspecting and finishing ties.

CHENEY BROTHERS 169

Plan to ship orders as nearly as possible complete, but do not unduly delay packing or shipment for a few special ties. Mr. Howell Cheney suggests this, and it seems the reasonable thing to do.

As we get better control of the stock and can plan manufacture better, the number of orders that are shipped incomplete should be rapidly reduced.

Scarfs:

  Schedule looms to make ribbons.
  Schedule broad goods to print and dye works.

REPORT ON VARIOUS MILLS

South Manchester, Connecticut
September 18, 1913

Weaving Mill:

  Try out new balance card for plain loom weaving. Start new man on Lamont's work.

Spinning Mill:

  Continue bonus work on doubling. Prepare for study on controlling.

Piece Dye:

  Study straightening threads the first time goods are dried, to see if they will not come through straight. Try Mycock Expander.

Folding-room:

  Rearrange work for preparing, ticketing, and invoicing. Make rack for sorting goods into manufacturing orders for checking.

Ribbon Mill:

  Have Miller complete the control of ties by records, also arrange to file partially filled orders by *destination* and *date*.
  Get two additional men at work in ribbon mill.

Yarn Warehouse:

Work up storeroom, stock-balance cards, system of handling orders and schedules. Decide amounts to be kept in stock, etc.

Winding-room:

Rearrange winding-room and build bridges.
Lay out work in receiving-room in order wanted.
Plan to arrange winders for bonus work.
Get man for order of work and planning, this man to be capable of making studies for task-setting when we get to that point.

Timekeeping:

Extend timekeeping.

Gantt asked for the same kind of direct, clear writing on the part of his associates as he was in the habit of dictating himself. Gordon Lee, one of his younger men who for a while acted as his secretary, recalls the following incident:

> One Friday morning he called me into his office. On his desk, as usual, was the week's mail, with a carbon copy of my reply attached to each letter. "Mistuh Lee, this letter of yours seems to be a little ambiguous. Read it again, please, and tell me what you meant."
>
> While I expounded it to him I could see that a different interpretation was possible, though unlikely, and said as much. In fact, I ventured the reassuring opinion that certainly Mr. H—— would understand it.
>
> He reached for the correspondence and perused it again. "Yes," he agreed, with judicial solemnity, "Mr. H—— is a highly intelligent man; I expect he will." He leaned back in his chair and gazed at me with his familiar droll scrutiny. "But, Lee, that isn't the point. Never write a letter that almost anyone can understand. No. Write every letter"—he waggled a thick, emphatic finger—"write every letter so that a damned

CHENEY BROTHERS

fool *can't misunderstand* it." He paused to let this sink in. "All right," he concluded, cheerfully. "That's all."

Some of the Cheney reports present in vigorous language Gantt's point of view in regard to superintendents, to the effect that they should accept the responsibility for their work and see to it that the work was done. Excerpts showing this opinion are these:

Report of June 9, 1915

Investigating and Task Setting:

The superintendent of each mill should keep closely in touch with the work being done by the investigators and task-setters in his mill. He should plan out and schedule in writing the work to be done by each investigator. By this means he will become familiar with the methods employed and the results obtained.

Inasmuch as the superintendent must assume the responsibility for the work of his mill, it is absolutely necessary that he understand in detail the methods employed.

Manufacturing Offices:

It is the function of the manufacturing office to see that the methods of manufacture which have been approved are carried out, and to bring to the attention of the superintendent, or his representative, all cases where this rule is violated, as well as all cases of bad or imperfect work.

Superintendent:

The work of the superintendent is, as far as the mill is concerned, thus divided into two classes:

The determination of methods

The carrying out of the methods adopted

The foremen are the direct representatives of the superintendent in the mill and should help perform both of the above functions. It is seldom, however, that we can get a foreman who can develop new methods and maintain those that have been developed, and he must usually be supplemented by the manufacturing office or the investigating department, as the case may require.

System of Management:

It is a mistake to assume that a system of management will lighten the responsibilities of the superintendent. A proper system of management brings to his notice things that are going wrong and gives him an opportunity to correct them. In this he must have the thorough cooperation of his foremen and investigators, who can materially lighten his work if they have the proper ability and training. There is usually plenty of such work to be done, but as the various operations are studied and tasks established the amount diminishes, and the investigators under the direction of the superintendent can be used largely for development and improvement.

Foremen:

When a system for the operation of a room has been established, the foreman must see that the room is operated in that manner. If he is incapable of doing so without frequently calling upon the man who put in the system, or is unwilling to do so, he is not fitted for the job and should be removed.

Report of January 26, 1916

Superintendents:

Of next importance is the fact that the superintendents should take more responsibility for the operation of the methods that have been installed. The greatest two obstacles in the way of their doing this, are, first, that they are already overloaded with work, and, second, that but few of the foremen really understand that it is necessary to obey orders.

No system can be operated successfully unless orders are

CHENEY BROTHERS 173

obeyed implicitly, and it is the function of the superintendents to train their foremen to *implicit obedience*.

I recognize that this is difficult in many cases; it is the one thing which more than any other is delaying the progress of our work.

Turning now to the first of the major developments in Gantt's methods which took place during the Cheney installation, consideration must be given to a paper which he presented to the American Society of Mechanical Engineers at the spring meeting in Buffalo in June, 1915. The title was "The Relation between Production and Costs." In this paper he stated a general principle in regard to indirect factory expense which was new and revolutionary at that time, but has since become a part of good cost-accounting practice. In his words: *"The indirect expense chargeable to the output of a factory bears the same ratio to the indirect expense necessary to run the factory at normal capacity, as the output in question bears to the normal output of the factory."*

It appears that he had given most careful consideration to this matter for some little time, studying it intensively, and trying out his ideas in practice, as was his wont. Edward B. Passano, president, the Williams and Wilkins Company, one of his early clients, showed him a report, dated January 23, 1913, on the possible increase in the percentage of net profit with an increase in the volume of business. From this report costs could be predetermined at any level of business activity. Gantt recognized at once the possibilities in such an attack on the problem of production costs, and pointed out to Passano the losses due to idleness when operating below plant capacity.

As a part of the argument in his Buffalo paper Gantt vigorously attacked the prevailing method of distributing overhead expenses or burden, saying:

> As a matter of fact it seems that the attempt to make a product bear the expense of plant not needed for its production is one of the most serious defects in our industrial system today, and farther reaching than the differences between employers and employees.
>
> The view of costs so largely held, namely, that the *product of a factory, however small, must bear the total expense, however large,* is responsible for much of the confusion about costs and hence leads to unsound business policies.
>
> Most of the cost systems in use, and the theories on which they are based, have been devised by accountants for the benefit of financiers, whose aim has been to criticize the factory and to make it responsible for all the shortcomings of the business. In this they have succeeded admirably, largely because *the methods used are not so devised as to enable the superintendent to present his side of the case.*

And then in support of his new theory:

> If we accept the view that the article produced shall bear only that portion of the indirect expense needed to produce it, our costs will not only become lower, but relatively far more constant, for the most variable factor in the cost of an article under the usual system of accounting has been the "overhead," which has varied almost inversely as the amount of the product. This item becomes substantially constant if the "overhead" is figured on the normal capacity of the plant.
>
> Our theory of cost keeping is that *one of its prime functions is to enable the superintendent to know whether or not he is doing the work he is responsible for as economically as possible,* which function is ignored in the majority of the cost systems now in general use.

Illustrations which he used to present the existing situation, and then, in contrast, that which would prevail under his methods, are most happy bits of argumentation. His example of current cost practice in applying burden is this:

> As an illustration I may cite a case which recently came to my attention. A man found that his cost on a certain article was thirty cents. When he found that he could buy it for twenty-six cents, he gave orders to stop manufacturing and to buy it, saying he did not understand how his competitor could sell at that price. He seemed to realize that there was a flaw somewhere, but he could not locate it. I then asked him what his expense consisted of. His reply was, labor ten cents, material eight cents, and overhead twelve cents. My next question was, are you running your factory at full capacity? and got the reply that he was running it at less than half its capacity, possibly at one-third. The next question was, what would be the overhead on this article if your factory were running full? The reply was that it would be about five cents, hence the cost would be only twenty-three cents.
>
> The possibility that his competitor was running his factory full suggested itself at once as an explanation.
>
> The next question that suggested itself was how the twelve cents overhead, which was charged to this article, would be paid if the article was bought. The obvious answer was that it would have to be distributed over the product still being made, and would thereby increase its cost. In such a case it would probably be found that some other article was costing more than it could be bought for, and, if the same policy were pursued, the second article should be bought, which would cause the remaining product to bear a still higher expense rate.

His second illustration, which in his own words "seems to put the subject in its true light," is this:

> Let us suppose that a manufacturer owns three identical plants, of an economical operating size, manufacturing the

same article—one located in Albany, one in Buffalo, and one in Chicago—and that they are all running at their normal capacity and managed equally well. The amount of indirect expense per unit of product would be substantially the same in each of these factories, as would be the total cost. Now suppose that business suddenly falls off to one-third of its previous amount and that the manufacturer shuts down the plants in Albany and Buffalo, and continues to run the one in Chicago exactly as it has been run before. The product from the Chicago plant would have the same cost that it previously had, but the expense of carrying two idle factories might be so great as to take all the profits out of the business; in other words, the profit made from the Chicago plant might be offset entirely by the loss made by the Albany and Buffalo plants.

If these plants, instead of being in different cities, were located in the same city, a similar condition might also exist in which the expense of the two idle plants would be such a drain on the business that they would offset the profit made in the going plant.

Instead of considering these three factories to be in different parts of one city, they might be considered as being within the same yard, which would not change the conditions. Finally, we might consider that the walls between these factories were taken down and that the three factories were turned into one plant, the output of which has been reduced to one-third of its normal volume. Arguing as before, it would be proper to charge to this product only one-third of the indirect expense charged when the factory was running full.

At Cheney Brothers Gantt made a definite application of this principle of idleness expense. In his report of July 26, 1916, he wrote: "Satisfactory progress is being made in regard to idleness expense and we hope to have it going in all the mills by October 1."

On September 26 of this same year he referred to the

progress being made in regard to the study of idleness, and specifically referred to idleness expense charts. The excerpt is:

> I feel that the points at which progress can be made are best shown by the *Idle Expense Charts.*
>
> We have these charts for nearly all the mills now, and hope soon to have them for every activity in the works.
>
> These charts, if carefully studied, will help us to determine policies of management, which are vastly more important than efficiencies of workmen. I cannot emphasize this fact too much, for unless we get rid of the expense of idleness as far as possible, it may counteract the beneficial effects of the most efficient operation.

Gantt repeatedly recommended, and rather insisted upon using, 100-per-cent activity as normal, and charging the entire amount of idleness computed on this basis to profit and loss. Cheney Brothers deviated slightly from this proposed procedure by setting up normals in the various mill departments, somewhat lower than 100 per cent, and charging a small portion of the cost of idleness into the product. Subsequent management practice in general has adopted 85 per cent of maximum capacity as the base for computing factory idleness.

In the spring of 1918 Gantt testified before the Federal Trade Commission in a hearing with reference to fixing the price of newsprint paper. Here he presented in considerable detail his theory of separating the costs of idleness from the costs of production. A somewhat extended excerpt shows the development of his power of presenting his beliefs in regard to idle expense over that of his paper of 1915. It is an excellent example of his condensed, vigorous, crystal-clear style, with here and there a flash of colloquialism.

Q. Will you state in a general way of just what your work consists?

A. Consulting with reference to industrial management.

Q. Do you have anything to do with cost-accounting problems?

A. The cost-keeping problem almost invariably comes up in one form or another when we have begun to study the production problem.

Q. The purpose of your employment is usually for advice to assist in promoting efficiency, reducing costs, increasing profits—is that a fair statement?

A. No. I should say my work is primarily engaged in increasing production. All of the others follow, if the production is properly increased.

Q. When you say increasing production, is that with the idea of building new plants, or new equipment, or increasing production with what they have?

A. Increasing production with what they have.

Q. Have you in connection with your work had occasion to consider the matter of accounting costs—the elements that enter into manufacturing costs?

A. Yes, I have.

Q. And those elements are labor, materials, and expenses, are they not?

A. Yes. Some people call it burden or overhead.

Q. Upon what basis, in your opinion, should those elements of cost be taken into consideration in determining the manufacturing cost?

A. I don't think it is a question of opinion. I think it is a question of fact and can be demonstrated. Cost is a question of fact rather than a question of opinion. The direct labor in operation daily, and the expense incurred for material, can be ascertained; the question of burden or overhead is the bone of contention. How much burden or overhead may be chargeable to any particular operation is the one thing which people disagree on and which apparently makes the most of this discussion.

Q. When you suggest that there can be no difference as to materials, you say materials should be taken in at actual cost? That is, the original cost, the carrying charge and expenses, everything ready to go into the factory when it is wanted?

A. Yes, and whatever interest you may have had to pay.

Q. That is the cost—actual cost—of the material?

A. I define cost as the actual expense incurred.

Q. You would not, then, charge your materials in at the market value if the market value was greater than the actual cost?

A. I don't think we have any right to do that, because we should be charging an increment which we have nothing to do with producing.

Q. When you come to the question of expense or burden, the difference of opinion that you have suggested arises largely from the factor of idleness, does it not?

A. Other things, too, but idleness is one.

Q. The question of idleness is always an important one with regard to expenses, is it not?

A. Yes.

Q. If your mill is entirely idle, your manufacturing plant entirely idle, you get no returns; you cannot charge anything to the consumer. If it is 50 per cent idle, what do you do?

A. We put the expense of maintaining that 50 per cent in a column by itself.

Q. Where does it ultimately go?

A. If I owned my own business, I should put it in profit and loss. Some people prefer to see how much of that they can charge back to the consumer, but those who do so in a competitive business will very soon have no idle expenses at all.

Commissioner Colver: Why?

A. Because the man who has the lowest cost and can afford to sell at the lowest price will gradually fill up his plant, while the man who insists on charging for his idleness will have a higher cost of manufacture and thereby be cut out of the market.

Commissioner Murdock: I wonder if that works in the flour-milling industry. Now, the flour mills of the United States cannot grind to capacity because they have not got the wheat to grind to capacity. Wheat does not exist, so they have got to have the factor of idleness there. Do you suppose they make a separate charge?

A. I am afraid they don't. I am afraid they charge it all to the public.

Q. They sure do.

A. In a competitive business, however, that mill which charges the least to the public will get the most business. Now we have this in operation, and it is working, and the plants that have adopted this method are getting ahead of their competitors, so it is not a question of theory any longer. It is a question of fact.

Commissioner Colver: Coalless Monday would not come under that rule? There isn't any idleness that can be cured by any voluntary act.

A. That is a question. If a plant is forced to suspend operations for the benefit of a community at large, it must seek its return from the community at large, and not from the particular customers that use its products. It may have redress against the government, but not against the people that use the product. They should not be charged with it.

Q. With respect to the treatment of investments in manufacturing enterprises, how should they be considered for the purpose of figuring depreciation and rates of return? Should investments be taken—physical properties—on the basis of a replacement cost at the present time, or at the original cost?

CHENEY BROTHERS

A. I should like, before I answer that question, to have a little more concrete definition of cost—what is cost?

Q. Suppose you define that term as you use it?

A. Perhaps I might, with your permission, explain some of the things we have run into. I have been interested in the cost question for only the last four years, about. Previous to that time I found that costs were very much as anybody chose to make them. No two cost accountants seemed to have the same ideas. When I have worked for concerns I have not discussed much with them how they should figure their burden because some of it was satisfactory to me. When the war came on in 1914 and business dropped to one-quarter of what it had been, everybody said his cost system was no good and he had to disregard it because if he used it it would make his product cost so much he could not sell it. I had known that for some time, and that is what I did not like about it, because the thing did not work in an emergency and anything that will not work in an emergency is not very good. It is in the emergency we want the thing, and not when things are running smoothly. Some of us got busy to find out what was the real trouble, and we found that they had been charging—all the cost accountants had been charging—to the cost of an article all the expenses incurred during the progress of manufacture of that article, and asking for a profit on top of that. Secondly, costs varied up and down according to seasonal expenses. So after studying that matter out, we came to the conclusion that that theory was entirely wrong; that if the cost of an article included only those expenses needed to produce it, this cost would not fluctuate so much with variations of business; it would be very nearly stationary.

COMMISSIONER COLVER: Variation of business. You mean by that, volume?

A. Yes. Figuring the cost of typewriters, for instance, by the old theory made it impossible to sell them. Figured on the new theory, it went on all right. The competitor who tried to use the old theory and based his selling price on the cost figured that way, could not sell any. We found the same thing in other industries; if you charge to the cost of an article only those expenses needed to produce that article, you get a relatively stationary cost. There is a big difference between the term "cost" and the term "price." They many times have but little to do with each other.

COMMISSIONER MURDOCK: Usually they do not have a speaking acquaintance, from my observation.

COMMISSIONER COLVER: There are a lot of things which you exclude from what you call—perhaps what I call—true cost; a lot of things in factories which we ordinarily find in cost, which you exclude, but which finally are not actually rejected, but do find their way back into price, or somewhere else, thoroughly identified, thoroughly known, and not at all repugnant to any good bookkeeping or good business. Is not that true?

A. Yes; but I want them under separate cover, and want them identified.

COMMISSIONER MURDOCK: What are they?

COMMISSIONER COLVER: I believe that a very great part of the disagreements that are coming about in all this price-fixing and cost-fighting and all that sort of thing, is only because we are not quarreling about letting something in, but we should like to know what it is when it goes in.

WITNESS: The drummer's silk hat, in the cost of his trip, for instance.

A specimen of the idleness-expense chart, a most important part in Gantt's development of graphical presentations, is reproduced on page 183. In the wide column headed "% of capacity used on day turn" the amount of capacity used is plotted by drawing a straight line beginning at the left and extending to the proper point.

| WORKING PERIOD 4 | WEEKS 22 | DAYS 192.0 | HOURS | PERIOD ENDING May 28 | 1927 | 78.5 % | Heating (Inc.) MILL |
| TOTAL MACHINE EXPENSE USED 56713.46 | TOTAL UNUSED 16112.21 | TOTAL CAPACITY USED 77.9 % | TOTAL CREDIT .042 % |

SYMBOL	DEPARTMENT OR MACH. CLASS	% CAPACITY TO ATTAIN	% OF CAPACITY USED ON ACT. TURN 10 20 30 40 50 60 70 80 90	LACK OF ORDERS	LACK OF HELP	LACK OF RAW MATERIAL	LACK OF WORKED MATERIAL	REPAIRS	POOR PLANNING	TOTAL	UNAVOIDABLE AMOUNT	%	EXPENSE OF IDLENESS INCREASE IN EXPENSE OF PRODUCT %	AVOIDABLE	REMARKS
B	200 - 45" Looms	80		F 47 32 2892 31	136 10		19 40	20 92	6 39	3122 44	868 31	20	18.2	2254 13	A - Poor Material 17.66
															B - No Help 342.76 Weavers Out 1144.63
C	697 - 54" & 56" Looms	80		F267 80 864 76	1473 82		16 54 194 14	493 86	136 42	3456 07	5646 65	20	18.2	2210 56	Wait for Fixer 345.08 Wait for Twister 973.88 " StopMotion 123.00 Doctor's Office 29.28
D	136-65" & 65" Looms	80		F142 60 432 85	441 94		1 28 5 65	41 53	6 30	1069 83	1161 66	20	18.2	20 10 81 83	C - Uph. Looms Running grey goods 450.27
E	9 - 56" Looms	70		F 6 55 376 15	24 93		4 65	14 34	38 99	463 67	207 02	30	36.0	256 65	D - No filling I.W.M. 347.77
F	111 - 68"&72" Looms	70		F194 66 1088 22	725 26		84 57 33 9	288 54	271 12	24 77 2888 35	8 92 3895 80	30	36.0	33 69 1007 42	E - No filling P.P. 331.02 F - Samples 676.56
G	1 - 72" Looms	80		0 52 82				1 98		54 78	9 92	20	18.2	44 86	G - Shut down
H	1 - 56" Loom	80		0 30 12	10 90				4 48	41 50	3 10	20	18.2	38 40	
J	26 -54" & 65" Looms	80		F291 57 23 58	40 67		6 20	94 28	21	456 21	203 96	20	18.2	252 25	
CC	39 - 54"&56" Looms	80		F 11 85 275 50 296 30	105 34		35 60	38 72	84 93	581 41	388 04	20	19.4	193 37	
	Storage			13 06						13 06				13 06	
				7101 81	2958 80		D 697 77	994 45	474 45 12127 32	12391 06			2062 72	
										67 16				2393 65	
										12060 16				350 93	

IDLENESS EXPENSE CHART

Gantt was so impressed by the importance of this chart, and by the discovery which it represented, that he gave instructions to have it patented. The necessary application was prepared by an attorney in Hartford, Connecticut, and made ready for Gantt's signature. Charts prepared at that time carried in the lower left-hand corner the legend, "Patent applied for." Before making a final decision to patent or copyright, Gantt, as was his custom, proceeded to think the situation through. He finally decided that he would not protect the chart, saying that as a management tool it was of too great value and importance to American industry to be controlled by one man. The application was dropped without having been filed and the chart was given freely to the world.

CHAPTER XIV

THE YEARS OF WAR

THE events of the World War had an absorbing interest for Gantt. From boyhood he had been a close student of military campaigns. He had a detailed knowledge of the strategy and tactics of the world's famous generals, and an intimate, almost technical, knowledge of the plan and action of every major battle of the Civil War.

From the background of this close study into the plans and prosecution of war it was but natural that, from the outbreak of hostilities in August, 1914, he should follow the movements of the armies, progress of the battles, defeats and victories, with an intensity and keen estimation of results, which would be expected only of one who had been educated and trained in military affairs. At home he had a large globe on which he picked out day by day the location of the battle lines. He also had large-scale maps on which he plotted with a soft pencil the location of the armies on the western European front. These maps were always up to date.

Gantt's study of military operations did more than fit him to follow and understand the ebb and flow of the struggle in Europe. In a very real sense, his great service to America in the World War was made possible by his previous researches into the conduct of wars. In particular, he seemed to draw upon his knowledge of the teachings and experiences of General "Stonewall" Jackson in planning his own work and pushing his outstanding contributions forward to a successful conclusion.

During his boyhood days at McDonogh, Jackson was his particular hero. In his mature years he ranked Jackson as "the greatest of all American generals." He continually advised his friends and associates to read Henderson's *Stonewall Jackson and the American Civil War,* and so learn of the masterful methods that contributed so much to the speed and success of his campaigns. A comment which he was fond of reiterating was to the effect that Jackson was notable for the clearness of his objectives, for careful planning, not only of the movement of troops, but for their supplies of munitions and food, and for the supporting work of the engineers corps. It was not only Jackson's military brilliancy which Gantt admired, but he likewise respected his character and appreciated his mental qualities. One of his favorite paragraphs from Henderson is this:

> Command over his [Jackson's] attention was formed into a habit which no tempest of confusion could disturb. His power of abstraction became unrivaled. His imagination was trained and invigorated until it became capable of grouping the most extensive and complex considerations. The power of his mind was drilled like the strength of an athlete, and his self-concentration became unsurpassed.

A direct contact with European events was afforded Gantt through his friend, James F. Butterworth, of the London firm which was the British agent for his piling-machine. The correspondence which passed between them during the first three years of the World War was devoted quite as much to information and comments in regard to the struggle as to business affairs.

The invasion of Belgium by the German armies stirred Gantt to the depths of his just nature:

> To my mind it was as distinct an act of savagery as could possibly be imagined, and it was all the more criminal because it was evidently planned beforehand.

His judgment of the flow of events was in many instances prophetic. On September 19, 1914, a little over a month after the first declarations of war, he expressed these two opinions, that the war would be long, and that its end would come from the breakdown of civilian Germany: "I am afraid the war will be a long one, but I do not believe the internal system of Germany will be able to stand the strain as long as the military organization will. I feel that defeat will ultimately come from internal collapse even before the armies succumb."

On December 5, 1914, he again turned to the question of the duration of the war, writing to Butterworth that "the first indication in the collapse of the German strength will be the fall of Austria," and "I do not think that Kitchener's estimate of three years was at all extravagant, and while many people in this country felt that the present situation could not last that long, I feel that they are beginning to realize the possibility more and more that Lord Kitchener was not far wrong."

The intensity of his own feelings at this time he reveals in a letter of March 19, 1915: "I personally feel so strongly with regard to the war that I can hardly trust myself to say anything on the subject. As far as the Germans are concerned, it seems to me they think a thing is right because they do it, and wrong because somebody else does it. They do not seem even yet to consider that there is a possibility that they may ultimately be worsted." His righteous anger flamed again in a letter written on the 10th of the following May:

> When Germany repudiated her most solemn agreement by invading Belgium, she put herself outside the pale of civilization, and many in this country recognized it, perhaps even more clearly than did the average Englishman. We hoped, however, that it was due to a temporary state of mind. We now find that state of mind to be chronic, and there seems nothing to do but to regard her as an OUTLAW which must be suppressed by the civilized world at any cost.
>
> I hope President Wilson will, at the earliest possible moment, call a conference of representatives from all the neutral nations to formulate a plan for doing away with this serious menace to civilization.
>
> In the meantime Congress should be called to ask for an appropriation of $500,000,000 to carry out the findings of such a council.

His predictive judgment of the situation was once more revealed strikingly in his belief that the English and French forces would not be able to cross the Rhine, and that Austria would not continue as an ally through the winter of 1916-17. This view he expressed to Butterworth in a letter of July 29, 1916: "I very much doubt if even the combined English and French forces will be able to cross the Rhine, but I have every expectation that Austria will be out of the running before another winter is passed."

The plight of the Belgian refugees, their misery and hardships, touched Gantt's ever-active sympathy. He sent a check to the Belgian Relief Headquarters in New York City. Acknowledgment of this contribution came through the office of J. P. Morgan and Company. He resented this banker's connection with the relief agency, exploded as was his wont on such occasions, and thereafter made his contributions to Belgian relief through London. In fact, he instructed Butterworth to turn over to the relief agency all of the sums which were owing to him from the business arrangement under his patents for the sale of British-built machines.

Early in 1916 the war brought a sorrow to him in the death of a cousin, a member of the Canadian forces, who was killed in the battle of Neuve Chapelle.

Up to the entrance of the United States into the World War Gantt had had little contact with governmental activities. In fact, there had been but two such occasions. In 1911, George von L. Meyer, Secretary of the Navy under President Taft, arranged for a study of the organization and management of the navy yards by a committee of experts. There were three members in this group, Gantt, Charles Day, and Harrington Emerson. These engineers rendered a report which was ineffectual so far as any practical results were concerned. Secretary Meyer announced in October of that same year that none of the systems of scientific management adopted in the United States would be used in any of the navy yards. Instead, he proposed to import from Europe what was known as the "Vickers System."

Gantt's second governmental connection was as consulting engineer for the Frankford Arsenal just before the United States entered the World War. Colonel George Montgomery, who was commanding officer at that time, gives these words of appreciation: "Mr. Gantt greatly impressed me as a lovable man who handled employees, who were naturally opposed to any form of shop management, in a very satisfactory way. He met with cordial cooperation everywhere throughout the plant. He pointed out errors of management and production, but always impressed the management that solution of resulting problems was the work of the officers of the Arsenal. He was such a gentleman on all occasions that it was a distinct pleasure to have him as an expert in shop management."

Colonel J. H. Pelot, who was in charge of the Artillery Ammunition Department at Frankford when Gantt's work was begun, and later became general superintendent of production of the entire Arsenal, recalls his democratic attitude, enthusiasm, and common sense: "For a man of his caliber, Mr. Gantt was one of the most democratic, most readily approached, and most companionable men with whom it has ever been my pleasure to be associated. His interest, enthusiasm, and methodical common-sense ways of handling production problems were simply contagious and were of much benefit to all of us who in those days of increasing production requirements for ammunition, were working at such top speed to produce the best results."

As was to be expected, Gantt was a keen observer of what was taking place in munition plants in the United States, in those factories which were endeavoring to turn out rifles and shells for the Allies. He sensed the causes of the difficulties and inefficiencies which were everywhere apparent and talked about them and wrote concerning them in words of warning. In an article in the *Engineering Magazine* for September, 1916, under the title "What Is Preparedness?" he made a vigorous attack upon managerial failures:

> The most casual investigation into the reasons why so many of the munition manufacturers have not made good, reveals the

fact that *their failure is due to lack of managerial ability* rather than to any other cause.

Without efficiency in management, efficiency of the workmen is useless even if it is possible to get it. With an efficient management there is but little difficulty in training the workmen to be efficient. I have proved this so many times and so clearly that there can be absolutely no doubt about it. Our most serious trouble is incompetency in high places. As long as that remains uncorrected no amount of efficiency in the workmen will avail very much.

. . . our industries are suffering from lack of competent managers—which is another way of saying that many of those who control our industries hold their positions not through their ability to accomplish results, but for some other reason. In other words, industrial control is too often based on favoritism or privilege rather than on ability. *This hampers the healthy, normal development of industrialism, which can only reach its highest development when equal opportunity is secured to all and when all reward is equitably proportioned to service rendered. In other words, when industry becomes democratic.*

In the same article he condemned autocratic financial control, saying that American industries were being managed by arbitrary power exercised by financial interests.

When the United States declared war on Germany on April 6, 1917, Gantt at once began to plan how he could best serve his beloved country. The contribution he made is a major one, although it has never been written down, is unknown except to a few of his close associates of those days, and has never received any public recognition. At no time in his career was Gantt's modesty more in evidence. He brought the greater number of his associates into responsible governmental positions to aid in carrying on the war, and in so doing he cut off nearly all of the income from his consulting practice.

His three great achievements were:

> Controlling the production of rifles, guns, ammunition and other war material for the army.
> Speeding the building of ships for the Emergency Fleet Corporation.
> Improving the operation of ships for the Shipping Board.

To accomplish these results he perfected the Gantt Chart as a managerial tool, selected "rivets driven" as the unit by which to measure progress in the building of ships, and picked the "ship hour" as the unit to measure the performance of ships. Any one of these accomplishments is a sufficient contribution for one man to make to earn the gratitude of his Government and praise of his people. Gantt trebly deserved such recognition.

On several occasions he was offered a commission in the United States army, but persistently refused. He accepted no compensation from the United States Government, not even reimbursement for his out-of-pocket expenses. He gave his energies, his resources, himself, unstintingly to America in her time of great need. One of his close friends, Edward B. Passano, who saw him frequently in action during these strenuous months of his life, comments upon his character as it matured under stress to become a reflection of the lives of his forefathers:

> To me the most interesting side of H. L. Gantt's character was developed through his work for the Government during the war. I was privileged during part of that period to work for him or under his direction. It was a peculiarly interesting experience to me in view of the fact that our previous relation-

ship had been just the reverse, he having served in a consulting capacity to the concern of which I happened to be the head.

During the war when I saw him in action under so many different conditions, the one thought which seemed to revert constantly to my mind was the influence of heredity in a man's life, what he actually owes to his progenitors, to what extent he is not only influenced by the character of his forebears, but how it would seem at times that in actuality he is in a sense living again in part their lives.

The work Gantt did for the Government during the war was with several agencies. First, the Ordnance Bureau of the United States army, later the War Industries Board, then the Emergency Fleet Corporation, Shipping Board, and in the production of naval aircraft. S. Marshall Evans, who was associated with Gantt during much of this work, pictures the great national need of the moment, saying that when it had reached a peak of intensity the Government finally turned to men who knew how to do: "They were there in Washington in considerable force. Vainly had they been pointing to the simple ways of hard work by which our vast requirements could be understood and procured. These men were those who had spent their lives in construction and production, and the greatest and most potent of them all were the engineers. Of the engineers who volunteered their services, many of us believe, *Mr. Gantt was the foremost.*" (The italics are the biographer's.)

General William Crozier, Chief of Ordnance in the early part of the war, in a letter to the author, dated May 20, 1920, told of the series of events which gave Gantt his opportunity to develop methods for the control of army ordnance material.

> I had previously made use of Mr. Gantt's services as an efficiency engineer, directing his employment at the Frankford Arsenal, and was pleased and encouraged with his methods and their promise at that establishment.
>
> As soon as we got into the World War I realized that the immense expansion of production would soon place us in possession of a quantity of finished articles and partially worked materials, and in a position with reference to manufacturing orders and deliveries, which would be utterly bewildering unless we could establish a better system for keeping track of our output and supply than had ever been found necessary in time of peace. I therefore sent for Mr. Gantt and asked him to assume the task of installing in the Department a method which would keep us continually informed as to our state of progress, so that we might at all times know where we stood and what we had to expect. I wanted him to accept a commission in the Department, but he thought he could do better work as a civilian, because he also expected to work for the assistance of certain civil agencies of the Government in providing for the war, and I deferred to his opinion.
>
> Mr. Gantt installed his system of straight-line charts for keeping track of production, and gave much assistance in securing personnel for handling them, some of whom were commissioned and some employed, and I relied upon them greatly for my own information and that of my principal officers.

Albert R. Brunker, president, the Liquid Carbonic Company, who first met Gantt in the Ordnance Bureau, and later became one of his clients, gives this further testimony as to the broad effectiveness of the methods Gantt installed.

> General Crozier told me personally that while all the systems [of management] seemed to have some merit, when he wanted data quickly on a particular subject and had to have facts, he always went to Gantt's work for the answer, thus clearly demonstrating that where real facts were needed, pre-

THE YEARS OF WAR

sented in a way that was simplicity itself, he was far in advance of all others of his group.

The task to which Gantt was assigned was coordinating the work of production not only in the Government arsenals, but in all of the privately owned plants working on ordnance material. The problem included keeping track of the progress of the multitudinous and extremely varied orders for artillery, small arms, machine guns, ammunition, explosives, etc. Into this situation Gantt threw himself without reserve. In a comparatively short time he had installed his simple method of charting the progress of production to enable those in charge to have a comprehensive, detailed picture not only of what was available and what had been accomplished, but what had to be done in order to deliver complete equipment to accompany the units of the American army as they were prepared for overseas.

His concentration on the industrial problems brought by the war was complete. Nothing else was in his mind. No other idea could get his attention. "I have walked down the middle of Pennsylvania Avenue in Washington at his side," says F. D. Manning, "while he talked on a problem. He apparently did not realize he was not on the sidewalk. I have lunched with him, and while he talked on work I have often been tempted to ask him suddenly just what it was he was eating. I have always felt that he would have had to stop to see what it was before answering."

Henry Harrison Suplee retains as one of his distinct recollections of the engineering work done to carry on the war, the thoroughness with which Gantt identified himself with the undertaking before him. He kept the ultimate result continually in view: "It was this power of concentration which particularly impressed me. During the strenuous summer of 1918 I remember him in Washington, a driving force in the great final effort preceding the Armistice. Nothing could divert him from his conviction of the truth and the directness of its enforcement."

The basic thinking upon which Gantt was acting at this time he revealed in a few paragraphs in an article published in the *Evening Mail* of Philadelphia, April 24, 1917:

> We believe that the greatest possible strength lies in the right kind of democracy. We are opposed to the ideas of those who believe that democracy is essentially weak, and in times of stress must be bolstered up by "patches of autocracy."
>
> We believe that to make democracy strong it must be more democratic. The new democracy does not consist in the privilege of doing as one pleases, whether it is right or wrong, but in each man's doing his part in the best way that can be devised from scientific knowledge and experience thus cooperating with all others.
>
> True democracy is attained only when men are endowed with authority in proportion to their ability to use it efficiently and their willingness to promote the public good. Such men are natural leaders whom all will follow.
>
> The idea which so many people have that in a democratic community every man may do as he pleases without regard to his neighbors, so long as he does not actually violate the law, is giving place to the higher ideal that men should so coordinate their activities as to benefit the community. Further, that any action which is detrimental to the community is wrong, whether or not it happens to fall under the ban of the law.

His great purpose in this war work was to get the facts, fix responsibility, decrease idleness, forward production. Naturally, he ran against tradition, and aroused jealousies and opposition. At the outset, neither the civilian employees

THE YEARS OF WAR

nor the officers of the Ordnance Bureau seemed to appreciate the true situation. Thus Gantt had difficulty in getting facts, and when figures and records were presented to him they were often unreliable.

Amid confusion, inefficiency, cross-purposes, and favoritism, it was quite natural that Gantt's methods should arouse the opposition of those who were not making good. The facts they revealed pointed directly, continuously, and inescapably at failures. Thus his methods would carry on until their disclosures reached some one in authority whose failure they would expose. Then efforts would be made to block and discredit him and his work. He felt all this keenly, often saying to his friends: "I am alone in my efforts. Everyone seems to be working against me." There were those, however, who were earnestly interested in accomplishment rather than in obtaining personal credit.

Among the latter was General John T. Thompson in charge of the design and production of all small arms and ammunition supplied to the United States army. He adopted the Gantt charts with enthusiasm. At the close of the war he received the Distinguished Service Medal "for exceptionally meritorious and conspicuous service as Chief of the Small Arms Division of the office of Chief of Ordnance." At the time this medal was awarded General Thompson sent a copy of the citation to Gantt, with this acknowledgment and generous appreciation:

> A large share in this reward for the accomplishment of a great war task is due to H. L. Gantt and his assistants. The Gantt general-control production chart was my compass.

It was the effectiveness of Gantt's methods in the Ordnance Bureau that brought about their adoption by other governmental agencies, including the Emergency Fleet Corporation. When he began this part of his work the situation was truly desperate. During 1917 German submarines sank 6,618,623 tons of shipping. The world production of new ships during that same period was 2,703,345 tons, of which Great Britain and the United States, combined, constructed 2,100,000 tons. That is, during the first year the United States was in the World War the destruction was two and a half times the new building. This record, which alarmed Great Britain, France, and her Allies, was far from showing the gravity of the situation. To the sinkings there had to be added the ships which were damaged, only to limp into port and in many cases to become a total loss. Losses due to hazards of the sea were increasing because of the necessity of taking long chances, such as running without lights, operating with untrained crews, and disregarding the dangers of bad weather.

The simple truth is that when the Emergency Fleet Corporation sent for Gantt, Germany was paralyzing Allied shipping and so winning the war. Great Britain was in danger of starvation. As early as April, 1917, United States Ambassador Walter H. Page reported confidentially to President Wilson that the food in Great Britain was sufficient to feed the civilian population for only six weeks to two months. It was feared that by the end of the year Great Britain might be compelled to surrender because of lack of food. In this situation it was evident that America must build ships. To build ships the civilian population needed not only to be aroused, but to have some means of judging from day to day what was being accomplished. In this emergency Gantt realized that the first step was to select a unit which could be understood by everybody and which would, on the one hand, set the task to be accomplished, and on the other determine from day to day what was being done. His trained mind, supported by his detailed knowledge of industrial methods, selected "rivets driven" as this unit.

The attempts to arouse the people of the United States are well remembered by all who lived through those years. Posters, slogans, newspaper display, public gatherings, made the shipyard worker the hero of the moment. Work in the yards was made equivalent to service under arms. Shipyard workers were exempt from the draft, and their families were given service flags. Mass meetings were held across the country from the Atlantic to the Pacific. The public was aroused to aid the effort to build ships. The fabricated ship was developed and planned for mass production. Records of riveting crews were reported daily through the newspapers, with intense rivalry between various yards and crews.

The record shows one of the greatest outpouring of creative energy that the world has ever known. On the day of the Armistice the United States had 341 shipyards, 350,000 shipyard workers, and had had 1,284 launchings. On July 4, 1918, ninety-five ships were launched as the great event in the celebration of Independence Day. During 1918 the Emergency Fleet Corporation built and delivered 533 ships of a total of 3,030,406 tons.

Gantt's "rivets driven" set the task, measured the performance, and stated the terms in which the rival yards strove for records, and the American people watched the progress of the work.

After the ships were built they had to be operated. This

task was turned over to the Shipping Board. The situation at the time Gantt was called in to assist has been vividly described by Wallace Clark:

> The handling of this large and ever-growing fleet was a stupendous task—probably the most difficult problem which had ever arisen in the shipping world.
>
> It was found impossible at first even to keep track of the movements of vessels in general, to say nothing of determining whether they were on the right jobs and doing their work efficiently. For a time there was little progress. The old plan of tracing ships by sticking pins and flags on large maps was tried, but it was soon discovered that this system, with its thousands of pins and flags, was so cumbersome that it was impossible to follow the movements of even coastwise vessels. The most serious limitation of this system was that it did not take any account of time—a flag bearing the name of a steamer and stuck in a port gave no information as to how long the steamer had been there or where it had been before that.
>
> Card records were next tried, but there was such a mass of information and it was so difficult to secure any comprehensive idea of its tendencies or to visualize what was happening, that the information remained buried in the files.

Gantt saw clearly the problem of ship operation, realizing that ship efficiency must be striven for in the great emergency. He was anxious to arouse public interest in shipping-matters beyond the size of the fleet. He wanted to have this interest reflex on the minds of shipping-executives so that the latter would realize the public was judging them for the effectiveness of ship operation as well as for the number of ships under their control. Here again it was necessary to find and popularize a unit. He selected the "ship hour"; in his own words:

THE YEARS OF WAR 201

First, there should be popularized the idea of the "ship hour" as a basis of accountability of each and every ship—for the ship hour is independent of varying control, size, type, and cost per hour of different ships. On the one hand, ship operators are responsible for their vessels, "ship hours," and on the other hand, the owners of the fleet, whether private or Government, should be made to be interested in the performance of ships, ports, or fleets in terms of "ship hours." The American people, the real owners of the American fleet, should be led to have at least a sporting interest in the performance of their ships.

As a method he worked out a simple procedure for visualizing what the ships were doing day by day by means of ship-movement charts, a form of the Gantt Chart.

Dean Herman Schneider, who was close to Gantt's work for the Shipping Board, characterizes the situation in regard to handling ships in port as one of the major vexing problems of the war. The solution arrived at by Gantt was an achievement of the "highest order." To quote:

In many cases returning ships were in our harbor for as many as four weeks. Docking facilities were scarce. There was great congestion at the docks and the loss of time meant really a loss of moving capacity. Gantt undertook this work with zest and in a relatively short time had the stay in the harbor reduced to less than two weeks, including coaling, repairs, and loading. It was a remarkably fine piece of work, quickly and skillfully done, and Gantt received many compliments from the major officials on his performance.

Edward B. Passano, soon after the close of the war, when incidents were fresh and vivid in his mind, wrote an appreciation of the work Gantt did for the Shipping Board, Emergency Fleet Corporation, and in the production of naval aircraft:

At one time we well remember the country was clamoring for positive information as to when the first ships being built would be available for service. The Shipping Board and Emergency Fleet Corporation had been marvelously conceived and promptly created. Congress responded without delay to the demands of the President; the wealth of the nation in money, material, and men, was poured forth to meet the emergency of the situation. Still the devastation by the submarines to the shipping interests of the world was increasing rapidly.

The crying need was for ships, and the question was, when would they be completed? Numerous efforts were made to acquire reliable information as to what had been done and what could be accomplished in the future, but to no avail.

Gantt was called into consultation and his services retained. His training as a production engineer led him to establish a unit for measuring this class of production. He realized that it would be impracticable to determine results and set a task for accomplishment without first establishing and adopting such a unit. After analyzing the situation he offered the "rivets driven" as a means for measurement. It was found that practically the number of rivets was in direct ratio to the tonnage of a vessel; while not absolutely accurate, it was near enough for all practical purposes.

It was adopted, and very soon thereafter the country knew not only how far the work had progressed, but accurate estimates were made for future accomplishment. Even more important, those in charge were able to plan definitely. One of the results of this simple expedient which was not anticipated was the speeding up of production far beyond the plans and hopes of those responsible by the development of competition between the different yards throughout the country. The newspapers were full of it at the time. I do not know which yard or which gang finally held the high record; one week it was in the East, and the next in the West, and so on, but during this period production was increased.

Gantt's name was never mentioned in the newspapers in con-

THE YEARS OF WAR

nection with this work, and I do not imagine that one in ten thousand knows that he was the "power behind the gun," and that to him is due the credit for much that was done during the latter part of the war period in hurrying the completion of ships.

His method of charting used in connection with the work of the Emergency Fleet Corporation brought to light very shortly a condition which would have been humorous, it was so culpable, if the times had not been so serious.

After establishing the method of measuring the progress and production of building ships, he turned his attention to the equipment for them and soon found that contracts had been let for the building of hulls far in excess of the facilities of the country to manufacture boilers with which to equip them. In other words, there would have been numerous hulls completed, idly waiting for boilers and other equipment before they could be put into service. This illustration is only one of many showing to what extent Gantt's work was effective in overcoming obstacles and accomplishing results, yet he has not been credited to any extent in proportion to his service.

The application of his charting method to the handling of ships engaged in oversea trade proved very efficacious and was used extensively by the Allocation Division of the United States Shipping Board. It was something of a problem to keep a record of vessels plying between the United States and foreign ports, but it was done, and accomplished splendid results. The charts gave a graphic history or picture of each voyage, noting delays and the cause of such. They were the means of saving considerable time in the "turn around" and every day saved meant additional tonnage available for other service.

As soon as the charting of the performance of vessels envoyage was well under way, the problem of handling essential imports, such as nitrates, coffee, etc., was taken up by him and an estimate made of the total amount of each required by the country. The entering port for each commodity was established on the basis of its distribution upon arrival.

The charts showed the task to be performed and the performance from day to day, so that in cases of emergency, when extra vessels were necessary in some particular line of trade, definite information was available of arrivals, types of vessels, etc., and largely used in determining the proper allocation to make.

Out of this service grew a recording of performance of both ships while in port and docks engaged in oversea trade. From the study of lost time in port numerous special investigations were ordered made by the Government in such matters as loading coal, grain, and even general cargo. These investigations, to my personal knowledge, in several cases led to changes of method which largely increased the efficiency in handling, and the saving of much time lost.

Later, Gantt was called to the Quartermaster's Department, then to the Aircraft Division, at all times working far beyond his strength, without publicity, patiently persevering in training men to his method, reviewing their work from day to day, planning its future development and application, taking his knocks with a grin, from the incompetents and subalterns, and going ahead in another direction but keeping on.

His satisfaction was in accomplishment. He had no desire for the limelight, nor did he care for publicity or newspaper notoriety. To see him in his quiet, unobtrusive way going from department to department, encouraging this one, instructing the next one, planning for this department, and so on, you were brought to the realization that Gantt was "running true to form" as a patriot and a teacher, being a counterpart, you might say, to his pioneer ancestors in the building of our state.

Results like these do not come without skilled planning and the putting forth of tremendous effort. Gantt's work in Washington was actually a tactical campaign. This fact was continually in his mind, as well as in the thoughts of those who were working with him. Each time he arrived in Washington he would first make a study of the disposition of his forces. He investigated each group of men in the Ordnance Bureau, Navy, Shipping Board, and Emergency Fleet Corporation to see exactly where they stood, what problems faced them, what facts had been collected, what individuals outside of his own group could be counted on for support, and what others would be expected to oppose. He would then plan the next few moves in his campaign, taking great care to see that the facts were incontrovertible, and presented in such a way that they would be understood and accepted.

The aim of these efforts was to have munitions and supplies produced when they were wanted, and to provide the necessary ships to carry them to the Western front. Gantt's general method was to fix responsibility for taking the proper action at the right time, and then to let everyone know on whom this responsibility was placed. He worked quietly, with no publicity, but was continually meeting men in Washington who could be indirectly helpful. He had an influence on these men which they remembered for many years after the war was over.

Gantt had a remarkable power of concentration and a wonderful memory; he, therefore, could follow every movement in his campaigns, by direct contact while he was in Washington, and through letters and reports when he was away. He was always watching for openings into which he could advance his forces, and continually looking for possible opposition. He never exaggerated the dangers, but when he saw them approaching, he would always launch an offensive before the opposer was ready to attack. He lost many skirmishes, but his forces were so mobile, and he himself was so resourceful, that throughout his work in Washington he never received a major defeat to his plans. His ideas and methods were a major influence on America's industrial participation in the World War.

Chapter XV

THE GANTT CHART

J OE, come along with me. I have the whole world by the tail. I have got the best mechanism yet devised for controlling production and purchase programs." In this enthusiastic way Gantt told Professor Joseph W. Roe of the application of the Gantt Chart in war work. The time was the summer of 1917. The place was on F Street in Washington. After this outburst Gantt took Roe into one of the office buildings which had been commandeered by the Ordnance Bureau, and in a room on the upper floor exhibited a chart which was being worked upon by a young lieutenant. He realized even then the value and power of his new management mechanism, and with an intensity of manner and speech, which was a part of him, said: "We have all been wrong in scheduling on a basis of *quantities*. The essential element in the situation is *time*, and this should be the basis in laying out any program."

On April 19, 1917, Gantt reported to General William Crozier concerning the organization and management of the Ordnance Bureau and its manufacturing arsenals. The adoption of this report gave Gantt the opportunity to develop his graphical methods for presenting the facts in regard to needs for, and progress of orders on, all kinds of ordnance materials.

Gantt, throughout his entire professional work, had a

fondness for graphical presentations and the visualization of facts. It is possible that he was influenced by his knowledge of the usefulness of the blackboard in teaching. In any event, he developed his own graphical methods, step by step, and used charts and diagrams in his constant effort to simplify methods and procedures. He was continuously striving to compare performance with promise and often made a remark to this effect: "Let us set up our promises in the form of an ideal oil-painting. From time to time we will photograph our performance on translucent paper and lay it over our painting and learn in what respect and to what extent we are falling short of our promises. Then we can revise our promises as well as our plans for further performance."

To sketch the sequence of his development of graphics: The first step was his graphical daily balance chart developed at the American Locomotive Company as a part of his task-and-bonus system. Then followed the red-and-black bonus chart which was an outgrowth of his work for the Brighton Mills. The next step was the percentage chart which he used extensively at Saylesville. Then came a simple form of chart, used for both production and costs, where the main plotting was a straight line. This type was introduced largely in the factories of the Remington Typewriter Company. The final step, before developing the Gantt Chart in the form in which it has become widely known and used, was the idleness-expense chart which he originated at Cheney Brothers.

Examples of all of these charts are given in preceding pages of this biography. Taken together, they show a progressive development up to the Gantt Chart. This is in

THE GANTT CHART

reality an invention by itself, although Gantt had previously applied the horizontal straight line plotting in numerous instances. The kernel of the discovery is given in Gantt's comment to Roe with which this chapter opens: *The essential element in any performance is time, not quantity.* In this shift of emphasis Gantt was stressing the need and desirability of controlling rates, a philosophy which began to be accepted a decade after his death and has been crystallized into a law of management, which states:

> Operating performance is controlled most directly through control of the rates of expenditure of labor, materials, and expense (Alford and Hannum).

The invention of the Gantt Progress Chart antedates the World War. On page 78 of *Organizing for Work* Gantt writes: "Before the breaking out of the war a simple chart system, which showed the comparison between promises and performances, had been established in the Frankford Arsenal. This system General Crozier began to extend throughout the Ordnance Department as soon as we entered the war, in order that he might at all times see how each of his subordinates was performing the work assigned to him."

The first progress chart drawn in the artillery ammunition shops at Frankford Arsenal after the declaration of war was made by David B. Porter. It is reproduced on page 210. It was plotted before June 15, 1917, and shows data for the last week in May. The original is in Porter's possession and is probably the oldest Gantt Chart in existence. It is a progress chart on the manufacture of complete rounds of artillery ammunition. Twenty-eight items are listed, with data plotted for eight.

WEEKLY PROGRESS REPORT OF MANUFACTURE OF COMPLETE ROUNDS OF ARTILLERY AMMUNITION AT FRANKFORD ARSENAL

CALIBER AND TYPE	TOTAL NUMBER ON ORDER	WEEKLY SCHEDULE
1 - Pdr. Sub. Cal. Fixed	6000	
1 - Pdr. Sub. Cal. Unfixed	15,500	
1.7 - Pdr. Smoke Shell Type 3-B	5000	
1.7 - Pdr. Smoke Shell Type 4-A	15,500	
2.24" Shell	300	
2.95" Sea Coast Shot	6000	
2.95" Mt. Gun Shrapnel	63,577	1018
2.95" Mt. Gun Shell	23,841	3150
3" Howitzer Shell	1000	
3" Field Gun Common Shrap.	158,760	6664
3" Field Gun H.E. Shrapnel	60,568	
3" Field Gun Shell	325,076	5556
3" - 15-Pdr. Shell	35,000	
3" Shell Balloon Gun	300	
3" Shrap. Balloon Gun		
3.8" Howitzer Shrapnel	18,522	
3.8" Howitzer Shell	15,928	
4.7" Gun Shrapnel	228,597	
4.7" Howitzer Shrapnel	75,565	
4.7" Gun Shell	40,286	
4.7" Howitzer Shell	83,833	
4.7" Howitzer Shell Machine	2766	
6" Howitzer Shell	20,632	942
Rifle Grenade Live	138,248	1800
Rifle Grenade Dummy	3250	
Hand Grenade Live	127,600	
Hand Grenade Dummy	5425	1800
Illuminating Grenade	12,250	1800

First Gantt Chart Plotted for Artillery Ammunition

THE GANTT CHART

The only possible rival for the honor of priority in chart work is the Ordnance Bureau at Washington. F. D. Manning, who went to Washington to begin the making of charts of this kind, began his work sometime after May 17, 1917. He believes it was the last of the month. Although he cannot determine the precise date when his own work began, he gives the honor of priority to Frankford, writing: "It is my opinion that Frankford Arsenal had Gantt charts going both in the shell-shop and on small-arms ammunition, before any charts were started in Washington."

The first Gantt Chart to be published is reproduced on page 212 (*Industrial Management*, February, 1918). It shows the orders and output of five items of war material. The description which accompanied it gave this information.

The long line in each case carried forward to October indicates the accumulated needs as expressed in orders. The vertical figure at the end of each month gives the total requirements up to that date. The amount written horizontally in each monthly space is the amount to be supplied during that month. As plotted, the monthly divisions are equal in length. The amount indicated by this legend, however, varies according to the monthly needs. So the plotting is to a uniform scale as regards the element of time, and to a variable scale, month by month, as regards amount. Once the principle upon which the chart is constructed is comprehended, it is read at a glance and shows exactly the condition of the requirements and deliveries on the item under consideration. A glance at the first one plotted shows that the total requirements up to the end of October is 3,656,000, and the monthly requirements from November through October vary from 84,000 in December to 711,000 in June.

ITEM	UP TO NOV. 30	DEC.	JAN.	FEB.	MAR.	APR.	MAY	JUNE	JULY	AUG.	SEPT.	OCT.
REQUIREMENTS SCHEDULE	1906 M	1990 M	2074 M	2161 M	2251 M	2344 M	2443 M	3154 M	3273 N	3397 M	3525 M	3656 M
Ordered												
Completed	1906 M	84 M	84 M	87 M	90 M	93 M	99 M	711 M	119 M	124 M	128 M	131 M
Issued From Stores												
REQUIREMENTS SCHEDULE	2182 M	2342 M	2492 M	2647 M	2807 M	2972 M	3150 M	3948 M	4167 M	4392 M	4624 M	4862 M
Ordered												
Completed	2182 M	160 M	150 M	155 M	160 M	165 M	178 M	798 M	219 M	225 M	232 M	238 M
Issued From Stores												
REQUIREMENTS SCHEDULE	1997 M	2130 M	2266 M	2405 M	2547 M	2691 M	2844 M	3546 M	3737 M	3932 M	4130 M	4331 M
Ordered												
Completed	1997 M	133 M	136 M	139 M	142 M	144 M	153 M	702 M	191 M	195 M	198 M	201 M
Issued From Stores												
REQUIREMENTS SCHEDULE	268 M	281 M	295 M	309 M	323 M	337 M	352 M	446 M	464 M	483 M	502 M	521 M
Ordered												
Completed	268 M	13 M	14 M	14 M	14 M	14 M	15 M	94 M	18 M	19 M	19 M	19 M
Issued From Stores												
REQUIREMENTS SCHEDULE	986 M	1047 M	1097 M	1149 M	1202 M	1256 M	1313 M	1668 M	1739 M	1811 M	1885 M	1960 M
Ordered												
Completed	986 M	61 M	50 M	52 M	53 M	54 M	57 M	355 M	71 M	72 M	74 M	75 M
Issued from Stores												

First Gantt Chart to be Published

THE GANTT CHART 213

Comparing deliveries with needs on the first item, deliveries are a month ahead of requirements. Following down through the other items, it is seen that in every case the amount completed is ahead of the requisitions thus leaving a balance in stores.

Gantt applied these same graphical methods in all of his work during the war, and when the conflict ended and he returned to private practice he began their introduction at once into the factories of his clients. Thus the device had a development in his hands extending over a period of about two and one-half years. In keeping with his generous nature and sense of professional responsibility he gave specimens and explanations freely to all who asked for them. At the time of his death the following applications had received the most attention:

- To record the progress of work
- To record machine operation
- To record man performance
- To show the work ahead of a department or plant

While he was engaged in promoting the use of his charts he one day made this observation to an associate: "I have worked all my life and it is only now that I am doing the things I have wanted to do."

Subsequent work by Gantt's associates, and by many others throughout all industrial countries, have made applications of the plotting to all kinds of productive work. Preeminently it is the most useful of all management devices ever invented and developed. Wallace Clark says of it: "The Gantt Chart, because of its presentation of facts in their relation to time, is the most notable contribution to the art of management made in this generation."

The reason why it has had such widespread and universal adoption is because of the predominant advantages which it affords. Clark has summarized them in this way:

> The use of a Gantt Chart makes it necessary to have a plan. Recording that plan on a chart where it can be seen by others has a tendency to make it definite and accurate and to promote the assignment of clear-cut tasks to individuals. The plan is presented so clearly on these charts that it can be understood in detail and as a whole not only by the executive himself, but also by those above him and by his subordinates.
>
> The Gantt Chart compares what is done with what was done —it keeps the executive advised as to the progress made in the execution of his plan, and if the progress is not satisfactory it tells the reasons why. The executive's time is thus saved because each time a figure is received he does not need to compare it with past records and decide whether it is good or bad. He has determined once for all what figures will be satisfactory, and has recorded them on the chart. The comparison of the accomplishment with the plan then becomes merely a clerical task, and the executive is left free to study the tendencies and take the action indicated by the chart.
>
> The Gantt Chart emphasizes the reasons why performance falls short of the plan and thus fixes the responsibility for the success or failure of a plan. Causes and effects with their relation to time are brought out so clearly that it becomes possible for the executive to foresee future happenings with considerable accuracy.
>
> The Gantt Chart is, moreover, remarkably compact. Information can be concentrated on a single sheet which would require thirty-seven different sheets if shown on the usual type of curve charts. There is a continuity in the Gantt Chart which emphasizes any break in records or any lack of knowledge as to what has taken place.
>
> The Gantt Chart is easy to draw. No drafting experience is

THE GANTT CHART

necessary, for only straight lines are used. The principle is so simple that anyone with average intelligence can be trained to make these charts.

Gantt Charts are easy to read; no lines cross each other and all records move with time across the sheet from left to right. Charts drawn in pencil or black ink convey an impression of practicability, simplicity, economy, and strength which it is not possible to obtain by the use of colored inks or even squared paper. Since no colors need be used on Gantt Charts, prints are as intelligible and effective as originals.

The Gantt Chart visualizes the passing of time and thereby helps to reduce idleness and waste of time.

The Gantt Chart presents facts in their relation to time and is, therefore, dynamic. The chart itself becomes the moving force for action.

These quotations are from Clark's book, *The Gantt Chart*, published in the United States in 1922. Subsequently this volume was translated and published in French, Italian, Polish, Czechoslovakian, German, Spanish, Russian, and Japanese. An accompanying illustration shows the title pages of eight of these editions. The total circulation has been very large for a technical book. In Russia alone nineteen editions, aggregating about 100,000 copies, were published. The claim is made that when the Russian "Five-Year Plan" of industrial development was organized, it was completely plotted on Gantt charts.

The form in which Gantt presented his charting system to his clients is shown by the following excerpts. Most of them are from a letter dated May 27, 1919, to Cheney Brothers.

> To find out whether or not a factory is performing its function satisfactorily, we should ask the following questions:

TITLE PAGES OF VARIOUS EDITIONS OF "THE GANTT CHART"

THE GANTT CHART

1. Are all machines being used? If not, why not? See Machine Record Charts.
2. If running, are they doing the work most needed? If not, why not? See Progress Charts.
3. If running and doing the work most urgent, are they doing it as rapidly as they should? If not, why not? See Man Record Charts.

If all these questions have been answered in the affirmative, it is pretty safe to say that the factory is running in an entirely satisfactory manner which, of course, is indicated by the various charts above referred to. These charts are especially useful, however, in disclosing what is not going on in a satisfactory manner.

Machine Record Charts: These are used to show graphically when a machine runs and when it does not run, then the reason for its idleness. Inasmuch as the reason generally indicates who is responsible for this idleness, it is an incentive for the foreman to eliminate at once such idleness as he is personally responsible for. Moreover, it enables him to see at a glance not only how much of the idleness of his machines he is responsible for, but who is responsible for the remainder of it. Inasmuch as all agree that the machines were installed to produce goods, the foreman in whose office these charts are kept, not only advances his own interests by keeping them, but by the same means calls to the attention of all others their responsibilities with regard to keeping the shop busy. This information enables every man in authority to do his work better than would otherwise be possible.

Key to Machine Record Chart:

————	Time machine was working
▬▬▬▬	Cumulative time of individual machines
▰▰▰▰	Cumulative working time of a group of machines

The portion of the daily space through which no line is drawn represents the time the machine was idle.

Reasons for Idleness:

    H   Lack of help
    M Lack of, or trouble with, material
    O   Lack of orders
    P   Lack of power
    R   Repairs
    T   Lack of, or trouble with, tools
    V   Holiday

When there is more than one reason for idleness, the reason entered on the chart is determined by asking questions in the following order:

    Is the machine ready to run?
    Is there work for the machine?
    Are there tools?
    Is there power to run the machine?
    Is there an operator to run the machine?

Key to Progress Chart: At the left of the chart is a list of articles to be produced. The amounts for which orders have been placed are shown in the column headed "Amount Ordered." The dates between which deliveries are to be made are shown by angles. The amount to be delivered each month is shown by a figure at the left side of the space assigned to that month. The figures at the right of each space indicate the total due at that date.

If the amount due in any month is all received, a light line is drawn clear across the space representing that month. If only half the amount due is received, this line only goes halfway across. In general, the length of the light line indicates the amount delivered during that month.

The heavy line shows cumulatively the amount delivered up to the date of the last entry. It will be noted that if this line is drawn to the scale of the periods through which it passes,

THE GANTT CHART

the distance from the end of the line to the current date will represent the amount of time deliveries are behind or ahead of the schedule. It is thus seen that the short lines are the ones which require attention, as they are farthest behind schedule.

A, B, C, and D are summaries.

A is a summary of the orders shown on the lower part of the chart.

Man Record Charts: The record of a man's performance is comparatively valueless unless we are able to compare what he has done with what he should have done. Showing this comparison together with the reasons why a man has failed to accomplish the full amount, is of great help to any systematic attempt to remove obstacles which have prevented complete accomplishment.

The information as to how much a man should do for any kind of work in a given time, must be obtained and kept on cards which show the various operations to be done on all the parts which go through the shop, and how much should be done in an hour by a reasonably good man under normal conditions. The amount of work to be done in an hour should be determined by a capable mechanic or by the foreman. In any case it should have the full approval of the foreman. In this way his support in removing obstacles is secured, for he is interested in proving that his men can produce what he has said they can.

From the chart a foreman can see which man needs help most, and put his efforts where they are most effective. If a man is green, or lacks instructions on any particular kind of work, the foreman knows that either he or his instructors must give him more attention. If he is having trouble with material or tools the foreman must correct these troubles. If he is not given sufficient assistance, the foreman must provide it, etc. In other words, these charts make it possible to trace the lack of production to its sources, and in so far are a great help in removing obstacles. We have secured a remarkable degree of cooperation in this manner and have developed in workmen possibilities which had been unsuspected.

Nearly all workers welcome any assistance which may be given them by the foreman in removing obstacles which confront them, and are glad to be taught how to become better workers.

Key to Man Record Chart:

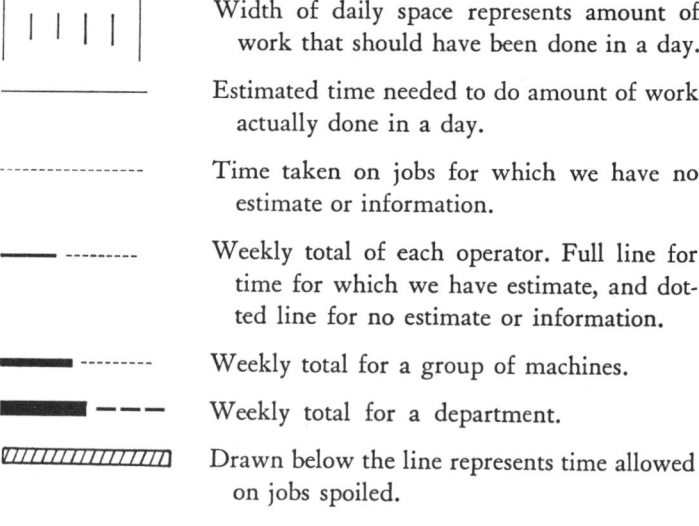

│ │ │ │ │	Width of daily space represents amount of work that should have been done in a day.
─────	Estimated time needed to do amount of work actually done in a day.
-------	Time taken on jobs for which we have no estimate or information.
─── -------	Weekly total of each operator. Full line for time for which we have estimate, and dotted line for no estimate or information.
▬▬ -------	Weekly total for a group of machines.
▬▬ ─ ─ ─	Weekly total for a department.
▨▨▨▨▨▨	Drawn below the line represents time allowed on jobs spoiled.

The portion of the daily space through which no line is drawn shows how much the man has fallen behind what was expected of him.

Reasons for Falling Behind

A Absent
B Transferred
C Careless, indifferent, or dissatisfied
D Defective work
E Machine stopped while being worked on by another operative
G Green operator

THE GANTT CHART

  H Lack of help
  I Lack of instructions
  M Lack of, or trouble with, materials
  O Lack of orders, or no work
  P Poor working conditions on account of lack of power, heat, light, etc.
  R Repairs needed
  T Lack of, or trouble with, tools
  V Holiday
  Y Smaller lot than estimate is based on

When there is more than one reason for failure to do the work in estimated time, the reason entered on chart is determined by asking questions in the following order:

Was the machine in good condition?
Were the tools and fixtures in good condition?
Was the operator given proper instructions and sufficient information?
Was trouble experienced with material?
On jobs requiring helpers, was help given promptly and properly?
Was the operator experienced and not too green to do the job?
Was lot smaller than estimate is based on?

These charts form a complete mechanism not for getting things done, but for furnishing the information needed by the executives in order to get things done.

David B. Porter has emphasized the universality of the Gantt Progress Chart, and of the way in which it has enlarged the possibilities of planning, in this statement:

> Although the Gantt Progress Chart was born in the factory, it soon outgrew the field of production and may now be found wherever it is possible to plan or schedule work and to measure its progress. The planning and execution of work underlie every

one of man's organized activities for his economic welfare. It is this wide application of the Gantt Progress Chart which has made it one of the greatest management mechanisms of the twentieth century. Just as the invention of the clock enlarged man's capacity to carry out a group enterprise by making possible a finer degree of coordinated activity, the progress chart has still further enlarged it by providing a means for planning in advance and then initiating an intricate series of interdependent activities, and for measuring the progress of fulfillment. The chart makes possible the visualizing of a multitude of events expressed in a universal language, and thereby extends man's control over industrial and commercial enterprise.

As an instance of an application outside of industry, its use in the police stations of Upper Silesia is of more than passing interest. J. Zoltaszek, chief inspector of the Upper Silesia police force, after a year of experience, gave this testimony: "They [Gantt charts] have so far entirely met their purpose, making the control, and as a consequence the issuing of orders, easier."

The nature of the application is indicated by the key to the lines representing the kinds of service done by a patrolman:

o	round
k	control
d	investigation
e	escort
kp	traveling control
p	emergency (reserve)
dz	office service
i	other service

If a policeman had no service on a certain day, lines are similarly drawn, and a system of letter symbols indicates the reason for the absence of the officer from regular duty.

THE GANTT CHART

 ur leave
 ch illness
 od rest
 dl detachment
 r miscellaneous

As to the technique of plotting, the police inspector says: "In factories every foreman knows how to draw them [Gantt charts] and I found that an average police sergeant handles them easily."

The record of uses and applications, after nearly a score of years, shows how prophetic was Gantt's evaluation of his own invention as spoken on that hot summer day in 1917.

Chapter XVI

PINE ISLAND FARM

ONE of Gantt's engineering jobs had sharp contrasts with all of his others. On this one he was his own client, working for himself instead of for another. The economic results were poor, widely different from the savings, earnings, and other substantial improvements which came from most of his efforts. The work done, instead of having in it elements of drudgery, was quite delightful; in fact, much of it was play to him. Instead of being carried on within the four walls of an office or factory, it was done under that sunshine which contributes so much to the gay spirits which are associated with the South. This self-imposed part of Gantt's engineering work, which meant so much to him in happiness and well-being, was the development and carrying on of his farm in northeastern South Carolina.

In one aspect the farm was his hobby. The contact with it and its problems carried him back in memory to his father's plantation and to the farm work which he did at McDonogh. It gave him opportunity for out-of-door life which was his only recreation. For formal games he had no liking, but there was a tremendous appeal for him in such sports and activities as hunting, fishing, crabbing, riding, driving, and walking. During his strenuous years at Saylesville it is remembered of him that he would always try to get away on Saturday afternoons for a half-day of fishing.

PINE ISLAND FARM

His fondness for this sport did not seem to be lessened by any kind of a happening, not even by the seasickness which he usually suffered when he went deep-sea fishing.

Perhaps, because of these likings, or due possibly to the pull of the soil from which he came, perhaps also because he was convinced that engineering methods could be applied successfully to the problems of agriculture, or it may have been from a desire to get away occasionally from the intense, perplexing problems of industrial life, in 1910 he bought a tract of land in South Carolina. It was peculiarly his own, personal project, for no other member of his family had any interest in or real liking for it and its surroundings. It was located in Horry County, near the town of Conway, about three miles from the seacoast at Myrtle Beach. It was reached by a branch of the Atlantic Coast Line Railways. A railroad station on the property was called Burcol, and this same name was given to a local post-office. The farm itself had been known as Pine Island. Gantt retained the name of the farm, and at his initiative the names of the railroad station and of the post-office were changed to correspond.

Pine Island farm contained 673 acres of land, about one-half of which was brought under cultivation, another 60 acres was good forest, and 80 acres more consisted of drainable swamp.

The section of South Carolina which is near the seacoast is made up of much swamp land, with higher ridges covered with long-leaf pine. Pine Island Farm had these characteristics. It had sandy-loam soil on the higher levels, accompanied here and there with some heavier fertile loam. The other sections were largely of clay soil underlain with a hard pan four or five inches thick, found some twelve or fifteen inches from the surface. Inasmuch as most of the cultivated areas were low the land had to be drained. For this purpose it was subdivided into plots each containing approximately four or five acres from which lateral ditches drained into main ditches. The latter emptied into a canal which was approximately three feet wide at the bottom and about five feet deep.

Certain definite factors seemed to have directed Gantt's attention to the possibilities of farming, and to the east-coast area. Prior to 1910 he had been doing work for phosphate mines in Florida and had become acquainted with the section around Sanford, where, by the aid of a system of underground drainage, many early crops were being raised, particularly large quantities of celery. One of Gantt's associates at Saylesville, George A. Dornin, had been for a number of years on a farm in Virginia. On several occasions Gantt had visited him in his Southern home. Discussions and interchange of points of view during these contacts indicated that truck-farming in sections of the South where the season was about two weeks earlier than the great truck section around Norfolk, Virginia, should be able to supply produce to the New York market at a time when it would be absorbed at satisfactory prices.

In general, Gantt decided to buy a farm a large part of which would be suitable for raising garden truck, but one which was in a run-down condition and so could be purchased without too great an expenditure of capital. Dornin carried out the investigation to find a suitable location and tract. He investigated a number of farms on a list which was submitted through the Agricultural Department of the Atlantic Coast Line at Wilmington, North Carolina.

Pine Island Farm was the selection. The area had originally been the site of a lumber camp with a sawmill and a turpentine-still. Tenant farmers had cleared the land. At the time of the purchase probably two hundred and fifty acres were under cultivation, and fifty acres more could be brought under tillage at slight expense. The larger proportion of the areas which appeared to be cultivatable had grown up to weeds and brush. Amid these surroundings Gantt started in to do scientific farming, or what might be better referred to as farming under scientific control. Dornin joined him as manager under an agreement for a monthly salary and a share in the profits.

One of the first steps was to make an intensive study of the quotations for various vegetables on the New York market for the years 1907 to 1910, both inclusive. These statistics were taken from the files of daily newspapers, principally the *New York Journal of Commerce*. The selling prices were charted for each of the four years for a period of from three to five weeks, the day limits being the approximate time when the truck crops would be ready for shipment from Pine Island. Through the use of an average of the prices over the four-year period, and an average of the middle of the contemplated shipping period, an average mean selling price was established as the income unit for the farm operations. Against this was charged the cost of raising the crop—labor, seed, fertilizer, container, packing, freight, and commission—to give the net profit per unit. Estimates were also made for profit per acre. These charts were made up for peas, string beans, Irish potatoes, sweet potatoes, cucumbers, beets, lima beans, and strawberries. From a careful study of these charts the crops were decided upon for the first year. In the spring of 1911 there were planted fifty acres of Irish potatoes, twenty of peas, twenty of string beans, ten to fifteen of cucumbers, ten of lima beans, and five of strawberries.

The land required heavy fertilization. The fertilizer ingredients were bought in bulk and mixed at the farm in the proper proportions for the different crops. This method saved the profit of the mixing-plant and reduced the freight charges.

These steps indicate the care with which Gantt preplanned his farm operations. A similar practice yielded methods for determining costs. Mention has been made of the way in which the farm was subdivided into four- to five-acre plots. Each plot was designated by a symbol composed of a letter and a number. Each designation was painted on a board about eighteen inches long and fifteen inches wide, nailed to a post about five feet high set at the northeastern corner of the plot. At noon each day worksheets were made up for the following day. The form carried the name of the workman, the name of the mule, plot number and symbol for the work to be done, and estimated time necessary for the operation. On some classes of work a man and mule would cover four or five plots in a single day.

As additional helps and sources of information, Gantt kept in close touch with the United States Department of Agriculture at Washington, and with the local agricultural authorities in South Carolina. He read continuously books and magazines on farm topics. Each time he visited Pine Island he would bring along two or three new books. These purchases gradually accumulated into a sizable library on crops, livestock, and agricultural methods. He required that a maximum and minimum temperature for each twenty-four hours should be observed and recorded, together with data on each day's weather conditions. For rainfall records he relied upon statistics from the Weather Bureau stations at Charleston and Wilmington.

The first year's crops, those of 1911, were small, although there was a fair yield of peas and string beans. A governing reason was a dry, hot spring. Following through a feature in Gantt's original plan, near the end of the growing season for truck crops, cotton and corn were planted in the middle of every other row to take advantage of the fertilizer which had been put in. These staples gave a fairly good yield. However, there was no net profit on the operations of that year. In fact, there was no net profit during any year. Some crops would come through as well as expected, but the farming on the whole resulted in a loss. Gantt frequently joked about this situation, saying that the farm was his "steam yacht."

In satisfaction it paid well. He made full use of the opportunity afforded for experimentation. At various times he raised loganberries, blackberries, raspberries, strawberries, casaba melons, honeydew melons, lettuce, romaine, endive, French artichokes, soy beans, vetch, watermelons, bur clover, alfalfa, chicory, and dasheen. The last named is a subtropical plant grown principally in South America and used in place of potatoes. The leaves resemble those of the elephant ear. The root is round, with eyes similar to a potato. When cooked, the inside is dark in color and tastes something like a mixture of potatoes, carrots, and chestnuts. Gantt intended to use them principally for feed for hogs. A few were shipped to New York, but there was no market demand for them.

Small quantities of all these fruits and vegetables were shipped to Montclair. Most of them were enjoyed, but no one seemed to appreciate the dasheen. One of his particular delights was to have served at home a dinner all of whose materials and ingredients came from Pine Island. On such occasions, chicory was substituted for coffee.

He was especially fond of watermelons, and generally had four or five acres planted, the seed being put in at different times, so as to produce a continuous crop. These were not raised with the thought of sending any to market, but rather to have plenty of melons for his personal consumption both at the farm and in his Northern home. When they were in season he much preferred to eat watermelons to drinking the Pine Island water, which tasted of sulphur and soda.

The truck crop in the spring of 1912, Gantt's second year, failed, due to too much wet weather, which drowned out and rotted the potatoes and other plantings. Dornin became discouraged and left. His place was taken by Steuart Jackson, one of Gantt's nephews.

Mention has been made of the local railway station and express office. Jackson became postmaster, express agent, and manager of the small general country store, or commissary. These activities indicate the extent to which Gantt was "in business." On the place were a number of tenant farmers, both white and colored, who raised staple crops. The general management of their affairs and the keeping of the accounts with them added to the details of farm operation.

In addition to experimenting with vegetables and fruits, truck crops, Gantt also did some work with staples. He endeavored to develop a strain of cotton especially suited to the conditions on his farm. The seed was planted in different types of soil and careful records were kept of the yield. Sufficient seed had been purchased so that he had quite an acreage in each particular variety of seed. On his visits to the farm he would spend hours at a time going through the cotton-fields, studying the various types of plants, numbers of bolls to the plant, and, as the bolls opened, the length of the staple. He would have a ball of cord with him, and wherever he saw an exceptional plant, or one he thought could be used for seed, he would tie a piece of string to it. The pickers were very particular to keep the cotton from those plants in a separate basket. Later on the seeds were carefully pulled out by hand, to be used in experiments the following year. Extensive tests were made to determine the germinating qualities of this seed, as well as of the regular seed from cotton that had been ginned.

To field corn Gantt gave the same study and thought as to cotton seed. Different varieties of corn were purchased, plants were carefully studied as they grew. When the corn matured he would tie a string around certain stalks he wanted kept for seed. The ears from these stalks were all set aside, properly marked, and given the seed-germination test before planting. Any that proved to be low in germination were thrown out. It was impossible to make this test on all the seed corn needed for the farm, but he tried to do it with an amount sufficient to plant five to ten acres. His purpose was to build up the yield.

Through one of his tenant farmers he became interested

in tobacco, and built a fair-sized tobacco-barn for the purpose of curing the leaves by artificial heat.

Gantt did not overlook the possibilities of livestock. He was keenly interested in raising hogs and bought thoroughbred animals for breeding. He tried Tamworth hogs, a large, red species developed in the South, and Berkshire. He raised pigs from the animals that were purchased and figured that he was on the way to making some money from hogs, when cholera attacked the herd. Anti-cholera serum and every known method of prevention were used, but nearly all died. This did not discourage Gantt and he kept on experimenting with hogs from time to time, but he never got to a point of making any profit on this investment. He was also interested in fattening steers. A part of the farm was fenced off, various cover crops put in, and more corn grown to feed them. It appeared that Southern cattle do not take kindly to being fenced in and properly fed; while they did eat plenty, the net result was practically *nil*. Many of the cows and steers were infected with cattle tick, which sapped their strength and seemed to keep them from growing large. Gantt put in a dipping-vat and had a deal of fun dipping the cattle, but even this process did not seem to help much.

As would be expected, he purchased the newest equipment so far as planters and plows were concerned. The potato-planter was so fast that none of the people on the farm could keep it from clogging and fill the hopper fast enough to keep it from skipping. Jackson had to run it. Gantt also purchased a disk plow to break up the soil twelve to fourteen inches deep. On much of the land it could not be used because of stumps.

PINE ISLAND FARM

The United States Department of Agriculture followed many of these experiments. On several occasions representatives of the Department visited the farm. The County Farm Demonstration Agent of Horry County was also interested in the new methods of operation. It was hard for local people to understand how so many acres could be cultivated with only ten mules. The secret was good management.

The farm brought other joys to Gantt than these absorbing experiments. He thoroughly enjoyed being out and about the whole day long with old clothes on. One of his especial interests was in the problem of proper drainage for the farm lands. He would spend hours working with a transit to establish the proper depths for the ditches. He liked nothing better than to go down into the ditch, mark off the depth and width which were to be dug, and then get back into the same ditch on his next trip to see the results of the deepening and widening. After spending a day in the ditches, coming out at night sopping wet and covered with mud, he would go into the farmhouse, take a bath, dress, eat his dinner, and sit around the rest of the evening discussing what he hoped to accomplish by the day's work.

In the evening when the weather was cool he liked to go to the store and talk to whomever was present. In the middle of the floor was a large wood-burning stove. He would stand close to this, his hands behind him, talking, commenting, joking, listening, and replying, seemingly finding as keen enjoyment in the flow of the remarks as did those who were talking with him. On other cool evenings he would sit before the open fire in the farmhouse, discussing some angle of the farm work. At such times he took a special delight in feeding the fire with fat turpentine wood, putting on a stick at a time and then watching the play of light and color from the darting, sputtering flames and the hot, glowing embers.

The cook on the farm was an old negro mammy, Aunt Judy. Gantt with utter solemnity would decide with her what he would like to have for his meals. Breakfast was usually waffles or corn bread. Mammy would get up early in the morning and start cooking waffles in order to have a supply ready for breakfast by the time Gantt arose. Usually she started too soon, with the result that the long-cooked waffles were soggy. Then Gantt would have to wait for a fresh lot to be made. Try as he would, he never could get her out of the habit of having plenty of waffles ready for the "boss man" for breakfast. Corn bread was one of mammy's specialties. On several occasions Gantt had Jackson write down the recipe for mammy's corn bread and send it North to him. However, he never succeeded in getting his wife, Mary, to admit she liked it as well as he did.

Pine Island Farm was on one side of a large, undeveloped territory. For this reason there was much wild life in the immediate vicinity. Gantt was fond of hunting in the late fall and winter. Quail were plentiful. He usually had one or two good hunting-dogs at the farm and would bring down Sam from Montclair. Often some one of his friends would come down with him to go shooting. During an afternoon he would bring in a bag of perhaps a half-dozen quail. At the time when cowpeas were ripe there were plenty of doves and he would occasionally shoot a mess of them.

The neighboring swamps were supposed to be good places for wild turkeys. Several times Gantt went turkey-hunting with an old negro guide. The guide usually brought back

PINE ISLAND FARM

one or two birds. Gantt never succeeded in shooting any. He always believed that the negro wanted two things, first, his money as a guide, and, second, the turkeys if any were shot.

Four or five miles northward above Myrtle Beach were several fresh-water ponds, a favorite place for ducks in the winter time. Occasionally Gantt went duck-hunting, usually with good luck. There was no fishing near Pine Island, so he could not indulge in that particular sport. He was, however, fond of driving around with an old white mule named Mike. But his real pleasure came from walking about the farm, studying his experiments, planning for new things to do, and giving general oversight to everything that was going on. Thus in spite of its engineering aspects, which were thought out with the same care which he gave to the problems of his clients, the farm was really a hobby. He often spoke the conviction that his health was decidedly benefited by the week or ten days he spent at Pine Island every three or four months.

The final record of Pine Island Farm reveals him as both pioneer and leader. When he bought the place, strawberries was the only crop produced in that part of South Carolina, except the staples, corn, cotton, and tobacco. Gantt, the newcomer, was the first to raise beans, potatoes, and other truck crops on a commercial scale. In so doing he set an example which was followed by his neighbors and their neighbors. Now (1934) the strawberries have disappeared, but the other truck crops are still cultivated and are Gantt's constructive, continuing contribution to the life of the entire section.

CHAPTER XVII

END OF A LIFE OF SERVICE

"I HAVE before me at least ten active years at full vigor, and then five more when I can take matters somewhat easier." In these words Gantt appraised his energy and looked toward the future, in the late autumn of 1919.

During the few months preceding this judgment of his own strength, he was planning for a reorganization of his professional practice with the purpose of taking members of his staff and some others into partnership. He was also working out certain changes and developments at Pine Island Farm to make that place a play spot for himself and his friends.

Early in 1919 he had published a little book under the title, *Organizing for Work*. It was packed full of vision as to coming events in the United States. A better name would have been *Organizing for Production*, but he had already used that title on a pamphlet published in the early months of the World War. A thought in his mind in the autumn of 1919 was to place alongside of this permanent contribution to the art and science of industrial operation a companion volume to be similarly and appropriately titled *Organizing for Distribution*.

None of these purposes, for his practice, his farm, and his book, was destined to be fulfilled.

One day, early in the third week of November, 1919, he had luncheon with some of his friends at his club in New

END OF A LIFE OF SERVICE 237

York City. Soon thereafter he was stricken with a digestive disturbance. He went home and sent for his physician. Medical aid could do little for him and he died on Saturday, November 23, 1919, at the age of fifty-eight years.

His remains were laid away in Mount Hebron Cemetery in Montclair, New Jersey. This resting-place is on a slope of the first Orange mountain, facing toward the east and the rising sun. His spot is marked by a plain, slate headstone upon which is a single biblical line selected by Mary Gantt as best fitting her husband's life and work: "I am among you as he who serveth." In simple truth, Gantt made the supreme sacrifice. He was a casualty in the cause of his country. The time of his passing away was but twelve and a half months after the signing of the Armistice that ended the World War.

Gantt's death was a shock to his friends. There was no foreshadowing of its coming. Within the short period of a week he passed from vigorous, robust health to his death. But his going, after there had been time for reflection, seemed to be a triumph for him. His life had matured and flowered during the three years immediately preceding. One and another, as they mentioned his ideals and his work, recalled lines from what was perhaps his favorite poem, Kipling's "When Earth's Last Picture Is Painted." The end of a life of struggle had come, and

> We shall rest, and, faith, we shall need it—lie down for an æon or two.

An opportunity for greater effort and higher achievement had opened, for he would now be like an artist who could

> . . . splash at a ten-league canvas with brushes of comets' hair.

His mastering love for eternal truth would now be satisfied for he would

> ... draw the Thing as he sees It for the God of Things as They are.

Letters of condolence and sympathy came to Mary Gantt from friends, associates, and clients, from every part of the United States and from abroad. They said in substance, "He was a great man, a master mind."

D. C. Lyle, Professor of Mathematics at McDonogh, wrote of the pride he had in Gantt, saying that of the thousand old McDonogh boys, no one was held in deeper affection or higher esteem:

> Only a few weeks ago he was here in this room, brimming with health, overflowing with energy and enthusiasm, and making me proud that he thought me worthy even to hear of his plans and his high hopes. Now he is cut off, his great work, his beneficent work, only half done. Though he was at the very top of his profession, his influence was growing daily, and there was great need of his sanity in the reorganization of economic conditions now going on. His loss is a national misfortune. Was there a more useful man in the country? I doubt it.

Harrington Emerson, an early exponent of efficiency engineering, one of the engineers who was with Gantt in the investigation of the organization and management of the United States navy-yards, elevated him to a prominent place among the great and famous men of the opening decades of the twentieth century. He recognized in him these outstanding characteristics and qualities:

> An energetic, dominant nature. Courageous and fearless in expressing his opinions, which he never gave without careful consideration.

END OF A LIFE OF SERVICE

Very decided when once his mind was made up.
Industrious and energetic in working out a problem.
Independent in thought and action, ready in speech and quick and impulsive in manner.
A clear thinker and a man of positive and decided character.
Precise and exact in his work and systematic in his activities, with a thoroughness in all his investigations worthy of imitation and praise.
He had the qualities of the leader in great measure:

> Good will toward mankind
> Ability
> Courage
> Charm of personality.

Frank and Lillian Gilbreth, his neighbors, admirers, and friends, paid a deserved, prophetic tribute to his leadership, expressed in these graceful lines:

> Courageously, loving Democracy
>   That gives each man a chance to be his best,
>   That makes for real cooperation, lest
> A new-won power lead to Autocracy,
> Steadfastly glad to stand the heavy strain
>   Of opposition and of the Slow to See
>   Knowing that finally the TRUTH MUST BE
> And peace and quiet come thru all the pain,—
> He preached the Gospel of real leadership,
>   In quiet words, with stress of facts and laws
>   Showing the goal and pointing out the way,
> Nor dreamed his words would found a Fellowship
>   Of those who held him Leader in a Cause,—
>   The winning of a new Industrial Day.

A year after Gantt's death, at the annual meeting of the American Society of Mechanical Engineers, a memorial session was held to commemorate his life and work. More than a thousand persons assembled in the main auditorium of the Engineering Societies' Building in New York City on the afternoon of Wednesday, December 8, 1920. Tributes were rendered to him as a man and as an engineer. The accomplishments of his professional career were appraised. Among the messages of appreciation was one from England and another from France. These showed how truly his influence had been international in its outreach.

James F. Butterworth, his English friend and business associate, believed that generations yet unborn would benefit from Gantt's teachings, and that "his services to the cause of humanity and efficiency will be an imperishable monument to his memory." Butterworth also referred to the appreciative references to his work which had appeared in British journals.

Fred J. Miller, writing from his long, intimate association with his friend, emphasized the fairness of his attitude toward all with whom he was associated. He himself played fair and insisted that those in whose services he was employed should do the same:

> His books teach above all other things, perhaps, the importance of the square deal in all that relates to industries or to the work of the industrial engineer; and no one, unless he has entirely lost faith in human nature, can doubt that in so doing he contributed in an important way to that which is and always will be a fundamental requirement for genuine success in industrial enterprises.

Gantt's remarkable work during the years of war was recalled by S. Marshall Evans. In 1918 President Wilson and Secretary of War Baker were bluntly told that by 1919 our allies would no longer be able to fight. The outstanding

The Henry Laurence Gantt Memorial Gold Medal

END OF A LIFE OF SERVICE 241

requirements from America were artillery, ammunition, airplanes, small arms, ships. The production of all of these was accelerated by the touch of Gantt's master hand. It seemed as if his whole life had been lived to fit him to meet this one great emergency of America. The soundness of the solutions he offered is evident to all who will study his record during those months of intense activity.

Charles de Freminville, writing for French engineering and industry, told of Gantt's friendliness to France and his understanding of the French industrial viewpoint, which is so different from the American:

> That which particularly gained the sympathies of the French engineers for Gantt was his constant endeavor to show that in order to successfully conduct organization work it is necessary above all to make certain the cooperation of all, and in order to do that, to make every one understand the purpose which is aimed.
> He understood and loved our country.
> He helped us powerfully.
> He will always inspire us.

Over and over again throughout that afternoon of tribute the thought was expressed that the memory of Gantt would not fade, that his work would live, and that his achievements would endure.

At about this same time Albert R. Brunker, who admired him and hoped to aid in perpetuating his methods and teachings, wrote: "Gantt stood for something far beyond the mere sordid desire to make money for himself, or the industries with which he became associated. His ideas made for the betterment of all classes, from the common workman up, throughout the entire range of business life."

R. A. Wentworth recalled his originality, idealism, and ability to work hard. "Gantt's outstanding characteristic was the effective manner in which he demonstrated originality and practical idealism in combination with the endurance needed to travel at night and work by day on the most practical of details. The high quality of his inspiration resulted mainly from incessant concentration on the problem at hand, whether in the application of science to an industrial process, in some refinement of administrative detail, or in the development of an advanced theory of human relations or economics."

William Eugene Pulis remembered Gantt's comradeship, the way he took young men into his counsels, the responsibility he placed upon them to make important decisions, and the inspiration of his life to help others to "carry on."

> To those closely associated with Mr. Gantt there was one phase of his leadership which endeared him to all, and that was his comradeship. To work with Gantt was a great deal more than working for him, and in this relation he placed a confidence in his men which called for the best that could be given. Fortunate were those who came in contact with this man of wonderful vision, ideals, and motives. Those who caught his spirit lost all thought of self and personal gain in an effort to reach the greater ideals for which he labored.
>
> When perplexing situations arose Gantt would gather all who were interested together and make each one feel that he had an important part in the solution of the problem at hand. After each gathering of this kind the influence of his personality on those of us who were present was an inspiration to return to the problem before us, and with a determination from which we could not be diverted accomplish the desired result. It was this power of creating in others a determination to master each problem as it came, that unquestionably made him a leader of men.

THE GANTT MEDAL

A MEMORIAL OF HENRY LAURENCE GANTT

ESTABLISHED TO HONOR

DISTINGUISHED ACHIEVEMENT IN MANAGEMENT

THE GOLD MEDAL HAS BEEN AWARDED TO

HENRY LAURENCE GANTT

for his humanizing influence upon industrial management and his invention of the Gantt chart.

BY THE GANTT MEDAL BOARD

L. P. Alford, CHAIRMAN   *W. J. Donald*, SECRETARY

MEMBERS OF THE BOARD MAKING THIS AWARD

INSTITUTE OF MANAGEMENT	THE AMERICAN SOCIETY OF MECHANICAL ENGINEERS
L. P. Alford	W. L. Conrad
Dwight Farnham	Dexter S. Kimball
Stanley P. Farwell	Conrad N. Lauer
Joseph W. Roe	

November 19, 1929.

CERTIFICATE ACCOMPANYING THE MEDAL AWARDED POSTHUMOUSLY TO GANTT

END OF A LIFE OF SERVICE

Gantt believed in forcing his young men to the front, and it was this practice which developed self-confidence in us. Many times after giving us his advice he would leave the solution of an important matter to one in whom he had confidence, and in most cases this responsibility was not misplaced. It would call for the best effort we could give, and in the event that the solution was beyond the person entrusted with it, wise counsel would be given and the credit bestowed for the effort. Gantt grasped every opportunity to give credit to others and it was this trait which inspired his men to be most loyal and unselfish in their work.

In business Gantt demanded honesty, justice, and industry, and he inspired all three. No matter how great the cost, honesty and justice had to be shown, and this lesson proved to be of the greatest value in carrying on his work.

Ten years after his death, on November 19, 1929, a group of his friends instituted the Henry Laurence Gantt Gold Memorial Medal. A fund was raised sufficient to defray the expense of an annual award, and the responsibility for selecting the recipient and making the presentation was placed jointly upon the American Society of Mechanical Engineers and the Institute of Management.

The first award was made posthumously to Gantt, the medal being presented to his daughter, Mrs. Margaret Gantt Taber. The inscription on the certificate reads, "For his humanizing influence upon industrial management and his invention of the Gantt Chart."

The Deed of Gift showing the purpose of the donors contains this declaration: "To memorialize the distinguished achievements and great service to the community rendered by Henry Laurence Gantt, management engineer, industrial leader, and humanitarian, we hereby found and establish in perpetuity the Henry Laurence Gantt gold medal to be awarded for *distinguished achievement in industrial management as a service to the community.*

The medal, designed by Kilenyi, shows on the obverse a portrait bust of Gantt executed with such skill by the artist as to be a striking likeness, and on the reverse the legend, For Distinguished Achievement in Management. This medal is regarded as one of the highest honors in the management field. Each presentation is made an occasion for reviewing and emphasizing the constructive philosophy of the man in whose honor it was established and is maintained.

Part III

Vision

FOREWORD

GANTT's creative powers of mind, his daring thinking, his mood to break with accepted beliefs, his constant search for what to do to progress—all of these traits and abilities are in truth shown by his work in the field of human engineering or socio-economics. A. N. Whitehead in his *Introduction to Mathematics* commends this kind of thinking. Such thought operations ". . . are like cavalry charges in a battle—they are strictly limited in number, they require fresh horses, and must only be made at decisive moments."

During the first decade of Gantt's professional work his goal in improving a client's industrial plant was to build "a self-perpetuating system of management based on the efficient utilization of scientific knowledge." In striving for this objective he became the unrelenting foe of arbitrary, autocratic domination, of incompetent control and management, of greed and avarice, whether of worker or owner or employer. He strove to develop by teaching and by practice truly democratic relations in business and in industry. He stood squarely for the right, which he defined in this way: "An action is right when it will advance the cause of humanity, and wrong when it will not." He believed in the rightness of the managerial mechanisms which he was installing.

As his experience deepened and widened he thought more and more of the social aspects and effects of industry in

America. Here he found as much to arouse his fighting spirit as he had met with in the operations of individuals. Characteristically, he began to search for a point of attack that promised to lead to a remedy. This attempt was abortive. Then came his work for his country during the World War. He learned much during those intense months. At their close he predicted economic disaster in the United States unless the actuating motive of business and industry should be changed. Facing the impending catastrophe like a seer of old, he exhorted his countrymen to substitute the motive of service to the community for the motive of personal profit that America might be strong.

Again, characteristically, he offered a mechanism, his system of visual records, measures of plan and performance, as a means to determine what motive is controlling action and what results are being achieved.

The record of these contributions to the pressing, and as yet unsolved, problem of human living is given in the chapters that follow, presented mainly in Gantt's own words. The excerpts are from many sources—professional papers, addresses, magazine and newspaper articles, and published books. There is considerable display throughout these quotations, obtained by the use of italic type and capitals. It is Gantt's own emphasis, and is a familiar feature in his writings.

Ample evidence can be presented to show the effect of Gantt's inspired leadership. As a specimen: During the decade following his death four economic engineering reports were issued by the American Engineering Council. Men who had been closely identified with him and with his work were on the committees that produced these progressive documents. The spirit of his beliefs and teachings is found in each one of these presentations of facts and declarations of principles. The report of 1921 on Waste in Industry, after listing the restrictions on productive effort, placed the major responsibility for improvement on the managerial group, as Gantt had done time and again, declaring,

> Over 50 per cent of the responsibility for these wastes can be placed at the door of management and less than 25 per cent at the door of labor. . . .

The Twelve-hour Shift Report of 1922 pointed out the influence of humanitarian considerations on decisions in industry. Gantt, preeminently among the engineers of his day, had advocated fair and just treatment of employees. In the words of the report:

> . . . the twelve-hour-shift day is too long when measured by twentieth century ideas as to the proper conduct of industry. Decisions are influenced today by humanitarian considerations as well as by the economic demand for that length of a day which will in the long run give maximum production.

The Safety and Production Report of 1928 reemphasized the responsibility of management for operating conditions:

> Industry is admirably efficient, its achievements are a source of justifiable national pride, but its processes must be freed from inexcusable human wastage. Its operation must no longer accumulate a presentable cost in human lives and curtailed energies. When these losses and costs have been eliminated or brought down to the irreducible minimum, then and then only will the highest productivity be secured and the most efficient operation realized. The dynamic force which can attain this highly desired objective resides in the management. Therefore,

a responsibility which cannot be evaded rests upon the management and executives of industry to make safety a major interest and continuing care.

Gantt had repeatedly stressed the possibility of increased production of goods, articles, and services. On one occasion he placed the output of American industry at only 15 to 20 per cent of what might be. The chapters on industry in the Report on Recent Economic Changes (1929) support his conclusions when they show that from 1919 to 1927 the productivity of the individual industrial worker had increased 53.5 per cent.

Truly, Gantt's pragmatic philosophy is a living force in American industry and engineering.

CHAPTER XVIII

DEMOCRACY IN INDUSTRY

Early in his management work, his speaking and writing, Gantt began to champion the principle of democracy in industry and to attack the practice of arbitrary, autocratic control. Management acts through organization. An organization requires sound, continuous direction. Direction comes from the executives. Executive decisions are determined by the attitudes and policies of managers and supervisors, as well as from facts.

Gantt declared again and again that control of industry should be in the hands of only those who "know what to do and how to do it." Decisions and actions should be based on ascertained facts, not on unsupported opinions. A scientific method of investigation should be used in seeking the solution of every business and industrial problem.

The soundness of these declarations was impressed upon him by the work for his clients, by his knowledge of the operation of American industry, and by his certain understanding of how progress in industrial performance was to be secured and maintained. Herein is one of his notable contributions to the industrial management movement in which he was an eminent, constructive leader.

Events subsequent to his death have shown that his vision of cooperation in industry, of the fair, just determination of every question between employer and employee on the basis of fact, of the elimination of all arbitrary acts and so of employer-employee warfare, was far ahead of the thought of his countrymen and countrywomen. In 1933, fourteen years after his demise, Congress, writing the will of the people into law, passed the National Recovery Act. Its celebrated Section 7 (a), which proved so difficult to interpret and apply, granted the right of collective bargaining to workers. This provision appears to be based on the theory that the interests of employers and employees are inherently antagonistic; that industrial relations must be fixed by bargaining, not determined by facts. One immediate effect of these terms of the Recovery Act was to release a flood of strikes in American industry.

The unlikeness between Gantt's purpose and this result is as wide as daylight and darkness. Gantt wanted and worked for industrial peace. He knew from extensive personal experience that if certain principles were adhered to, peace in industry could be attained and maintained. In contrast, the American people through their Government seemed to have turned away from cooperation and mutual understanding and to have promoted industrial strife.

Gantt's mature thought and teachings on democracy in industry are best revealed by excerpts from his numerous writings. They are direct, clear, and fearless in form of expression. Many are pungent and epigrammatic in wording.

> The truest definition of democracy is *Equality of Opportunity*.
> If democracy is to compete successfully with autocracy in the long run, it must develop organizing and executive methods which will be at least equal to those of autocracy.
> Our effort, then, is to approach as nearly as possible that ideal

DEMOCRACY IN INDUSTRY 253

community in which each man shall do the work for which he is best fitted and receive commensurate reward.

... any prosperity not based on efficiency is on an unstable foundation.

The most important problem before the industrial community today [1914] is that of the relation between the employer and employee. The reason why this is so is because the employer and employee do not understand each other.

... the era of force must give way to that of knowledge.

There is a general feeling that because our industries have in the past been directed in an autocratic manner, that autocracy will continue to be the rule, and that there is apparently no escape from it.

... the rapidity with which industrial development can be carried on by autocratic means is far greater than that which has so far been possible under democratic methods. On the other hand, the results obtained under democratic methods are far more permanent and less liable to be perverted to false ends. This leads us, therefore, to ask if autocracy in industry is not just as much a phase in industrial development, as we in this country consider it to be in political development.

Gantt was particularly vigorous in belittling the power of money and condemning the greed of corporations, and avaricious financiers and employers.

The great problem of the industrial leader is to solve the labor problem. The financier has assumed this task in the past, and the present deplorable conditions are the result. He has failed.

... *the idea so prevalent a few years ago in the industrial world that money was the most powerful factor, and that if we only had money enough, nothing else mattered very much, is beginning to lose force, for it is becoming clear that there is an end to the largest bank account, and that the size of the business is not so important as the policy by which it is directed.*

The Standard Oil Company, the Beef Trust, the Sugar Trust, and any number of others have absolutely no regard, apparently, for right or wrong. They get what they can by any means available. The difference between the savage and civilized communities is largely that the civilized communities have enacted laws which tend to restrain individual greed. Inasmuch, however, as it is impossible to foretell all the forms individual greed may take, it is impossible to enact in advance laws to cover all possible cases, and the best that can often be done is to make new laws to restrain new forms of greed as fast as they develop. Laws were made long ago that restrained robbers, sneak thieves, and even the "robber barons," but none have so far been framed that restrain the "high financier" who, without giving anything in return, taxes the community for his own benefit to an extent that makes all other forms of acquiring without giving any equitable return seem utterly insignificant.

His condemnation of wealth for itself led on to the declaration that men are of more importance than money:

> *Wealth is convenient, luxury is pleasant; but the nation which does not so develop its industries as to produce men, will not for any great length of time hold its place in the world.*
>
> It is imperative, therefore, in seeking the proper industrial methods to bear in mind the fact that the *men produced by them are far more important to the life and prosperity of a nation than the wealth and luxury by which we set so much store.*
>
> . . . that nation which first recognizes the fundamental fact *that production, and not money*, must be the aim of our economic system, will, other things being equal, exert a predominating influence on the civilization, which is to be built up in the period of reconstruction upon which we are now entering.
>
> Our immediate problem, then, is to develop a credit system that will enable us to take advantage of all the productive forces in the community. Such a credit system must not only be able to finance those who have ownership, but also those who have productive capacity, which is vastly more important. This is equivalent to saying that *our wealth in men is more important*

DEMOCRACY IN INDUSTRY 255

than our wealth in materials. So far we have never used this force to more than a small fraction of its capacity, simply for the reason, as previously stated, that the originators of our financial system were traders and not producers. Now, however, when the supreme importance of the producer has been recognized, we must enlarge our credit system in such a manner as to enable us to take full advantage of his possibilities; in other words, we must make it democratic.

Gantt recognized the antagonism between employers and employees which had led to the formation and governed the activities of labor unions. He condemned union methods as vigorously as he attacked those of financiers and employers.

> As long as the interests of the employer and employee seem antagonistic there will be conflict, and in any discussion of the subject we must recognize that antagonism means conflict.
>
> In discussing the relations between employer and employed, we must recognize the fact that in the majority of cases men still act on the principle that "they should take who have the power and they should keep who can."
>
> As a matter of fact, while the financier had been forming his great combinations of manufacturing interests and railroads, with the effect, at least, as far as the public is concerned, of upholding prices, the workmen had gone him one better. By their unions not only have they upheld the price of their labor, but in many cases markedly increased it, without rendering any more service than formerly; the employers, in many cases, say less.
>
> This is true whether you are speaking of employer or employed. Labor unions are just as insistent in their demands for things that do not belong to them, as the Sugar Trust is in its efforts to evade duties that it ought to pay. One of the best illustrations of this spirit of which I ever heard was incident to the ending of a strike in a Western State where the labor union had won. Soon after the men had gone back to work, one of the employers said to a workman, "I hope you are satisfied

now." "No," said he, "we are not satisfied, and we never shall be until we come to the works in our carriages and you walk!"

On account of the disregard of law and order that unions so frequently show in their strikes, it is the fashion in many places to condemn them as utterly bad, when they are only human. As a matter of fact, they are not all bad by any means. They have done a great deal for the cause of workmen. If it had not been for them, the working-people of today would probably be in the same condition as were those of England sixty or a hundred years ago. The average workman is a good citizen, just as loyal to his country as the capitalist, and just as proud of its position in the world. He is even more interested in its prosperity, for in times of depression, when the capitalist loses his surplus, the workman loses his means of living. It is a realization, perhaps, of the small margin that they have above their absolute needs, that makes workmen so liberal to each other, for it is a well-known fact that the wage-earner is far more liberal than the capitalist. He will go much farther out of his way to help a friend than the rich man will, although it is much harder for him to do so.

Men join labor unions from motives of self-interest. Gantt upheld their right to do so. The way to combat the union, he said, is to make it to the interest of men not to join. If this is attempted the plan must benefit the employer also.

Men join the union because they think they will be better off in the long run for being in the union. The idea of the union is to get a higher rate of wages for the whole class, because in general nobody in that class can get a substantially higher rate unless the whole class gets a higher rate.

There are many individuals who do what they can to help their less fortunate friends, and there may be unions formed to help the poor workman; but as a business proposition such a union cannot long be successful. Unions are formed, as a rule, by men of energy to help each other, and the poor workman is taken in, not for the good he does in the union, but the harm

DEMOCRACY IN INDUSTRY 257

he does if not in. The poor workman is thus advanced with the good, and the employer pays the bill.

> The desire of the union to take in all the members of its class is not philanthropic. Self-preservation is the first law of nature. Under ordinary conditions a man will advance himself first, and his neighbor next. He will join the union to advance his own interests, and it is only right and natural that he should advance his own interests. Any community made up of people who did not advance their own interests would very soon go to pieces. If a workman thinks it is to his interest to join a union, he has a legal right to do so. If we wish to prevent him, we must make it to his interest not to do so. In other words, we must provide him with means of advancing his interest that is superior to what the union offers. If any such scheme is to be permanently successful, it must be beneficial to the employer also.

His point of view in regard to employers' associations was very similar to that which he held regarding labor unions:

> Labor unions demanding all they can get, and employers' associations organized simply to oppose the demands of the unions, can never evolve a satisfactory system of management; for, although each, in its way, may be (and undoubtedly is) often beneficial to its members, both are formed with the idea of using force only, which can never be a substitute for knowledge.

He deprecated the lack of knowledge of industrial problems on the part of employers, and quoted with approval a statement to the effect that labor leaders involved in a certain strike had more intelligence than the employers against whom they were contending:

> A man who was sent by an independent set of employers to investigate the Lawrence strike, told me that he found much

more intelligence among the labor leaders than among the employers concerned, and that they had a far clearer comprehension of the problems involved. His mission in the investigation was to report to those who engaged him as to the best method of combating the I. W. W. They got the answer that nothing permanent could be done until the employers learned more about the industrial problems with which they had to deal.

He pictured the unhappy situation of the employer who is striving to obtain more than he pays for, and of the employee who is demanding wages for which he gives no adequate return:

> The employer who insists on more service than he pays for, and the employee who demands excessive wages for his work, both lose in the long run. The former worries continually about how to manage dissatisfied workmen who are continually on the verge of a strike, and in dull times the latter lives in constant dread that his employer may no longer be able to continue business and he may be out of work. In other words, unless efficient work goes with high wages, the result is apt to be disastrous to both employer and employee, and if we would have satisfactory workmen we must learn how to make their labor efficient, for it is to efficient labor only that high wages can be uniformly paid.

His point of view on the unemployed, or the idle, is that an idle class is a weakness in any nation:

> The war is making clear the fact that productive efficiency is the greatest force not only in industry, but in war, and hence *an idle class, whatever its excuse, is a serious handicap to any nation.*

He extended this thought of idleness, however, beyond individuals:

DEMOCRACY IN INDUSTRY

Idle capital is no more entitled to wages than idle labor.

... plants or people, therefore, who do not serve some useful purpose to a community are a handicap to that community, for idle plants represent idle capital, and idle people are not producers, but consumers only.

Gantt looked upon the setting up of democracy in industry as a step which would ultimately bring an abundance of the good things of life to every one in America. The possibility of the greater production of articles, goods, and services he was willing to express quantitatively:

> On the whole, only about 50 per cent of our industrial machines are actually operating during the time they are expected to operate; and on the whole these machines, during the time they are being operated, are producing only about 50 per cent of what they are expected to produce. This brings our productive result down to about one-fourth of what it might be if the machines were run all the time at their highest capacity.
>
> This conclusion is not a guess, but is based on reliable data. Unfortunately, there are many other elements of unnecessary waste in our productive process which cannot be so accurately calculated, *but which reduce our effectiveness certainly to 20, and very probably to 15 per cent.*

He maintained that democracy is strong, that it offers the opportunity for a release of creative energy, and that through it America can immeasurably increase her industrial efficiency.

> Our minds have been full of the beauties of democracy, but we have failed to appreciate its sublime strength. Democracy means the releasing of the infinite energy of all the people for creative work, something more than releasing their opinions for unlimited debate. I don't agree with the average politician's concept of democracy. His is the debating-society theory of Government; policies, according to him, must be de-

cided not by demonstrated facts but by opinions; not according to the laws of physics but by majority vote. Such democracy accomplishes nothing and leads nowhere. Real democracy consists of the organization of human affairs in harmony with natural laws, so that each individual shall have an equal opportunity to function at his highest possible capacity.

Such a democracy would be more efficient than any autocracy could possibly be. It would not be a tender plant appealing to sympathizers to "make the world safe" for it; it would be such a rugged growth that it would have nothing on earth to fear. Real democracy would not need a benevolent champion with a Big Stick to come to its support; and it would not need a sage with superhuman wisdom to tell it what to do. It would know what to do because it would be governed not by any one's opinion but by scientifically ascertained facts.

Voting upon a question of fact is the poorest possible way to settle it. Does water run uphill? This is an important question if we wish to install a water system, but intelligent democracy will not ask for a ballot on it. The best way, if we don't already know, is to find out. Let us do away with debating-society methods in both politics and industry and substitute demonstrated facts for the opinions of either an autocrat or the great majority. Until the people consent to such a change they will never attain to real democracy.

Today [1918] however, in both politics and industry the debating-society method is still in vogue. Labor troubles especially are seldom settled any other way. Employees ask for a certain wage; directors of the industry declare the demand unfair; then they either compromise or fight. The chances are that neither side has any data on the actual values concerned. The factory is probably controlled by financiers who know nothing of industrial management, while the labor leaders are unconcerned as to whether or not the work in dispute is of any value to the community. Strikes are called because it is presumed that they will pay. Lockouts are declared because the employers believe that it will pay to shut down the plant.

DEMOCRACY IN INDUSTRY 261

What sort of a system is this which places a premium on idleness?

He told how an engineer would eliminate the profit system, advocating a remedy which is the diametric opposite of many attempts which have been made in the name of recovery.

> An engineer's way of eliminating the profit system would be to encourage production and make the distribution of the product such a frictionless thing that salesmanship would become almost a lost art.
>
> We do not need to take over the factories into the hands of our universal debating society. The matter of their ownership would be a mere detail, anyway, if both production and distribution were scientifically managed. It is true that our great waste is caused by the system of production for profit; but it is not true that the lost wealth goes mostly to the profiteer. It doesn't go to anybody. It isn't being created.
>
> *Our problem should be to till our idle soil to the utmost possibility, to run our idle machines to their utmost capacity, and to make producers out of the vast armies of present-day busy people whose energies are being wasted in upholding our present inefficient system.*

The lot of the workers under industrial democracy would be vastly improved, Gantt asserted. The reason for this is that they would be working in their own interests. Every workingman and workingwoman would have full opportunity to earn all that he or she was capable of earning.

> Industrial democracy will look upon every worker not as an attachment to a machine, but as a unit in the organization. We do not need a revolution. We do not need a class war. Most people will work for the common good if you give them a chance. The trouble is that we have been clinging to an autocratic system under the mistaken notion that it at least was

good for the autocrat. The fact is that it isn't. Democracy is far better for all of us.

Sentimentally we believe in democracy, but we don't know what a great thing it is. What we need is not more laws, but more facts, and the whole problem will solve itself. . . . Such a democracy is the ultimate solution of the labor problem, for it settles the questions of wages, hours, and unemployment. The present system commits us to a vicious circle of strikes, higher wages, then higher prices, with a consequent reduction in the standard of living, which leads to further strikes. This is because production is carried on for profit and profits are determined not by industry in production but by competition in selling. By simplifying distribution, pooling it if you will, or socializing it—I am not concerned with the term—we will be able to arrive at the exact cost of production and will have an accurate and constant standard by which to determine the worker's worth.

As to hours, often the shorter workday has proved upon trial to be more efficient than the longer day; on the other hand, the worker would not be so concerned about hours if he were always working in his own interest.

In an indirect way, through attacking commonly held fallacies, he declared three doctrines of industry, or principles of management, which since his death have received considerable, though by no means universal, acceptance. These are the doctrines of

> High wages
> High supervisory expenses
> Low selling prices

The first of these fallacies, as Gantt expressed it, is: *"It is necessary to have low wages in order to have low costs."* This belief he held to be detrimental to all concerned.

The second fallacy is: *"In order to get low costs the expense of the supervising force must be small compared with*

DEMOCRACY IN INDUSTRY

that of those who are actually performing the physical work." He declared that, fortunately, this common belief was beginning to be discredited.

The third fallacy which he maintained was held by many business men, is: "High selling prices are necessary to large profits." In opposition to this belief he declared:

> *If we produce an article for which there is a large demand, and sell it for a price which most people can afford to pay, the cost of selling that article in large quantities will be extremely small.*

And, again

> . . . people are beginning to realize that there is no great ultimate profit in trying to sell a person something out of which he cannot get the value he paid for it.

His belief that a change was taking place from an autocratic control of industry to a more democratic system, is shown in this excerpt:

> . . . that our industries have been handled in general in an autocratic manner is no sign that they will continue to be so handled, and almost every day we see increasing symptoms that people are realizing what true democracy means.

And, finally, he expressed his disbelief in legislation as a cure for industrial difficulties, believing rather that the adoption of the scientific method was the way out:

> I do not feel, however, that the result is to come primarily through legislation, but by recognition on the part of employer and employee that there is a possible basis for mutual understanding, and that it is our duty to find it. Such a basis cannot be discovered by bodies of men resolving themselves into debating societies, but must be found by a thorough investigation by the scientific method of the industrial conditions as they exist.

Chapter XIX

THE NEW MACHINE

To "INCREASE the purchasing power of a day's work in New York City." This declaration of purpose outlines one of the first tasks set for itself by an organization in which Gantt was the moving spirit. It was given the name, "The New Machine." In the announcement of its organization appears this statement:

> It is intended that the New Machine shall be an organization for the acquirement of political as well as economic power and without any pretensions to higher motives than that of advancing the fortunes of its members by cheapening supplies and service to the elemental needs of the community. If it makes bread cheaper and the labor of man dearer the public won't complain.

During the middle months of 1916 Gantt had been reading writings of Thorstein Veblen, Professor of Mathematics at Princeton University, and Charles Ferguson, a special agent of the United States Department of Commerce. These men, Gantt said to his friends, although they are entirely outside of industry and have had no connection with its operation, have been able to diagnose accurately the primary causes of industrial ills. He made the acquaintanceship of Ferguson. It is not known that he met Veblen.

The trend of his thought, which received some energizing from these contacts, together with his knowledge of the

THE NEW MACHINE

inefficiency of industry, the laxity of business control, and the disturbing economic phenomenon of soaring prices of commodities, were the influences which prompted Gantt to make the move which led to The New Machine. The occasion was the annual meeting of the American Society of Mechanical Engineers held in New York City during the first week of December, 1916. On Tuesday of that week Ferguson published an open letter in the *Public Ledger* of Philadelphia which was a call for a union of masters of arts and of materials to operate business and industry for the benefit and advancement of the people:

> I say that the heart of modern politics—a regenerative politics that shall relieve the deadly moral and mental fatigue of decayed issues that smoke, but will not burn—lies in this ancient but still very new and exhilarating conception of a fighting university. I mean a conspiracy of men of science, engineers, chemists, land and sea tamers and general masters of arts and of materials—a fellowship at deadly enmity with all parasites and pretenders—held together in their war against humbugs by their common love of what is really so and their common scorn of purse-lipped, pious altruisms—a university that has escaped from the cloisters and flung out its gonfalon from the walls of the banks and the bourse.
>
> Yet most assuredly finance and industry must be socialized somehow. If we refuse to do it from the bottom we shall have to do it from the top, and doing it from the top means the emergence of many Prussias—with wars upon wars.

On Saturday of that week a news dispatch in the *Public Ledger* from a staff correspondent carried an account of the initiation of this movement, which was hailed as a crusade of fifty members of the American Society of Mechanical Engineers:

War was declared on inefficient managers of the great industries of the country by a schism of fifty members of the American Society of Mechanical Engineers at the annual meeting of the society, which closed its sessions today.

At two meetings that were not down on the program, and which their leaders expressly disclaimed as representing the American Society of Mechanical Engineers, the fifty members effected an organization, to be known as The New Machine, dedicated specifically to increase the purchasing power of a day's work in New York City.

The organization of this group, of which Gantt was the Chairman of the Executive Committee, was precipitated by two professional papers presented on Thursday afternoon of that same week. One was by Gantt under the title, "Productive Capacity a Measure of Value of an Industrial Property." A second was by Walter N. Polakov, "Valuation of Industrial Properties versus Valuations of Industrial Methods." Gantt insisted that the charges incident to over-evaluation of industrial properties increased overhead and unjustly raised costs. The direct consequence was a higher price of goods to the consumers.

> One thing which has been made clear by the war is that the most important asset which either a man or a nation can have is the ability to do things. For years with inefficiency at the top staring me in the face and hampering me at every turn I have labored to find a means of measuring that inefficiency; as it is perfectly evident that without efficient direction efficient workmen are ineffective, even if it is possible to get them, which it usually is not.
>
> If we can measure and evaluate the productive efficiency of the manager as we now measure that of the workmen, we may hope for better results.
>
> The only men organized for the promotion of productive

THE NEW MACHINE

efficiency are the engineers, and it is on your shoulders, gentlemen, that must fall the burden of showing what can be done.

Faulty management leading to the building of plants that are too large, maintaining improper and inadequate sales policies, over-estimating the demand for goods, adopting a hostile labor policy that results in lack of help at critical times, conducting buying and storekeeping systems that supply insufficient or poor material, permitting improper repair and maintenance of plants, and tolerating poor planning, the expenditures and losses from all these inefficiencies and insufficiencies are charged off by managers, Gantt declared, as though they were proper costs to be saddled upon the ultimate consumer. He accompanied this statement with a chart showing factors of managerial inefficiency traced in several industrial establishments in which he was installing new methods. Following the line of thought in regard to costs that he had developed somewhat earlier he added:

> The rule which must prevail in this country, if we are to meet the competition in trade and manufactures of the more highly organized industrial machines of Europe, is that the cost of an article shall include only those expenses actually needed for its production; any other expenses incurred by the producers for any reason whatever must be charged to some other account.
>
> The old rule that the cost of an article must include all the expense incurred in producing it, whether such expense actually contributed to the desired end or not, is a rule made for the convenience of idle and shiftless managers, but not for the service of the public.

Polakov's paper explored the ill effects of the misuse of the machinery and other factors of production which result in high production costs:

The solution offered by financiers and accountants has been to include in the manufacturing costs all the fixed charges of a plant and all the financial obligations of a company, irrespective of whether these investments or expenses contributed anything to the value of the commodity or in any way benefited the production.

The general misuse of or failure to use the expensive machinery of production is not a sufficient reason to either advance prices or to lower wages, as both determine our future. Evolution in our industrial relations will eventually reverse the present situation, and all charges for unprofitable investments and nonproductive forces will not be borne by the consumers, who as yet have no power or means to take an active part in the management of the industries. Unfit methods and incapable leaders are equally harmful to both investors and consumers, and their interests unite in demanding that methods of the highest value be developed and put into use.

The extent to which Gantt was the proponent and active force in this organization is shown by the fact that he put Ferguson on his payroll and made him virtually the manager of the movement. To separate this venture from his own business he used the Engineers' Club in New York City as the address for The New Machine.

What wishes or vision he had for this movement cannot be accurately determined. If he had dreams of a new political movement, or for the acquisition of political power by his group, these could have little, if any, reality in them, Gantt was the diametric opposite of the political leader. His inherent honesty, lack of tact, fearless straightforwardness, brusqueness of expression, and loneliness, because of lack of close friends, are not the human qualities which make for the success of the politician. Whatever Gantt's plan may have been for The New Machine, it never reached any degree of realization. The organization was kept alive for a few months, then disappeared soon after the United States entered the World War. During the short period of its existence it did hold a few meetings of its supporters, discussed some problems, and prepared and sent one letter to President Woodrow Wilson. Gantt's influence in the preparation of this communication can be clearly discerned, although some of the language appears to be Ferguson's. It condemns all forms of arbitrary power, and declares that America must not become regimented under Federal authority. It passes sentence upon irresponsible financiers and the advertisers' press. It submits that credit power and the free press are natural organs of social control. In an effort to make the purpose of the movement plain and reduce it to essential terms, a postscript was added from which the following paragraph is excerpted:

> What we mean is that the strength of the country, for peace or war, is involved in the operation of its business system. We put our technical knowledge and experience behind the proposition that the business system as it stands is ineffectual simply because it fails to simulate the creative forces. We have enough technology, but not enough liberty. We lack heart, sense, feeling.

For the reason that this document, although it was signed by nineteen men, is so unmistakably Gantt's thinking, both the letter and its postscript are reproduced in full:

LETTER TO THE PRESIDENT

THE ENGINEERS' CLUB, NEW YORK CITY
February 17, 1917

TO THE PRESIDENT OF THE UNITED STATES
WASHINGTON, D. C.

Dear Sir:

Pursuant to the suggestion contained in your communication of February 8, sent to Mr. Ferguson through Mr. Tumulty, the undersigned beg leave to describe the aim of a certain organization in which they are interested and to solicit your approval of it.

Our organization was formed last December at the Engineering Societies' Building in 39th Street, New York, by thirty-four men who were attending there the annual convention of The American Society of Mechanical Engineers. We have since increased considerably in numbers and now hold weekly meetings at this club. At a regular meeting today the subscribers hereto were authorized, as the executive committee of the organization, to make the following representation:

It appears to us, and to those for whom we speak, that the war which now engages the thoughts and emotions of mankind has come of a general failure to understand the social implications of the grand-scale productive process whose rise is the characteristic feature of the age. We think that no fact of contemporary politics can be understood until it is viewed in its relation to the dissolving and recreative power of this new-born world of work.

We believe that the vast and delicate mechanism of modern industrial society is so exacting in its demands upon the skill and probity of the race, so sensitive to the spirit of art and science, that it can gain and grow only through a progressive elimination of plutocracy and all other forms of arbitrary power.

Those of us whose business it is to advise concerning efficiency in factories, agree that the springs of high productive power are moral and spiritual; and that as the machinery becomes more intricate, the need becomes greater that there be nothing machine-like in the concord of the people that operate it. Thus it is through technical experience as well as by reflection that we have come to see that the ultimate power of tools—and therefore of arms—belongs not to tyranny, but to liberty.

We are therefore jealous and anxious that America shall be made invincibly strong through the rectification of its business system and the release of the creative faculties of the people. It is true, of course, that a people debilitated and disintegrated by the cold and senseless mechanism of plutocracy may gain a degree of strength by mere regimentation. But the great strength —the strength that can subdue all violence and enforce the peace of the world—can belong only to the nation that shall evoke the infinite resources of freedom—the inexhaustible reserves of imagination and enterprise—for the conquest of materials and natural forces and the mastery of the machines.

We think that the main outcome of the painful experience through which the world is now passing will be the closing of the gap between business and politics; everywhere business will be socialized and politics will be divested of abstractions and will engross itself in the struggle for economic strength; the idea of a business system working loose in a moral vacuum and devoid of social and scientific aims, and the idea of a politics devoted to subjective rights and careless of the earth-struggle, will in due time pass out of the mind and memory of the race.

We think there is no question about this interfusion of business and politics. The problem is not whether these two things shall become one thing—that is necessary and inescapable. The question is, which element shall prevail in the amalgam. Shall the old static order absorb the new dynamic order, shall the rigid conceptions of the historical state dominate the free and fluent constitution of the working world—or contrariwise?

We believe that Germany became comparatively strong among the nations of Europe, not because of its regimentation, but in spite of its regimentation and because of its scientific spirit and its investment of idealism in economics.

Nevertheless, the plain fact is that Europe has, for the present at least, closed the gap between business and politics by the sublimation of the State and the repression of the principles of business. Europe is Prussianizing itself—West and East as in the middle. European countries are thus made stronger than any plutocracy can be—stronger than any country can be that maintains the ancient schism and contradiction between unsocial business and uneconomic politics.

But up to the present moment Europe has, we submit, rendered substantially the wrong answer to the world's problem. It is left for America to give the right answer. It is left for America to demonstrate the incomparable power of an integrated society in which the free creative spirit of great business shall transform and mobilize the state.

The Prussianized societies of Europe can indeed work out within their several boundaries a better economy of men and materials than is possible to America in its present estate. But their external economics are beneath the level of science and civilization. By the very nature of their constitution they are made mutually repulsive. Their business is perforce belligerent in its external relations. They are fatally committed to a war after the war—and to wars upon wars. It is impossible to make a world-order out of this absorption of the earth-subduing powers by the state.

So it seems to us that the rôle of America is quite clear. We are to achieve an unexampled strength by resisting the temptation to Prussianize and by refusing to be regimented under an arbitrary authority at Washington. We look to Washington not for compressive force, but for releasing light and a coordinating intelligence. We ask you to use the massed credit and certificating power of the Presidency—the most availing moral and intellectual energy now extant in the world—to strike and destroy all arbitrariness and to captain the only good war: to wit, the fight of the artists and the engineers against every human thing that wilfully obstructs or opposes the creative process.

We mean to say that America alone can negotiate for the world the entrance to a new age, because no other country has such gifts of spontaneity and free constructive enterprise. No other country is in possession of a contagious principle of reconstruction that can cross all frontiers.

This is the mother-land of big business. Here only is it possible to conceive of a working order that is autonomous and that is lifted up to a stature of magnanimity and to a vision of self-vindicating law that can compass the world. It is possible for us to understand that the business system—the new social

THE NEW MACHINE

tissue formed through credit-capital, free contract, and corporate organization—is at bottom a civil polity that carries its own law within its own body, and that cannot harbor fraud, duress, or privilege without deadlock and self-destruction.

We say that business in the United States ought to be consciously and openly political—that the invisible government should be made visible. We think that people who earn their incomes—the great undertakers and organizers of industry, with the farmers and mechanics—ought to concert their political forces to free the shoulders of enterprise from the yoke of incomes that are unearned. We say this is necessary not for justice merely, but for something greater than justice—for creative, conserving and defensive power. We say it is impossible for the nation to be strong—that it will continue to be exposed to grave peril and will run the risk of utterly missing its destiny—if we do not find a way to take the control of the huge and delicate apparatus of industry out of the hands of idlers and wastrels and to deliver it over to those who understand its operations.

We do not dream, we offer no counsels of perfection. We are men of experience speaking of our own country and our own work, and we say that the industrial and commercial process in which America lives and moves has grown too complex to be run any longer by mere deskmen who sit as the agents of a leisure class. We say that for practical and directive purposes the tools must belong to those who know how to use them—not because it is more equitable that it should be so, but because it is impossible that it should long continue to be otherwise.

Yes, there is one other possibility. We can Prussianize—as England and France are doing. There is no other alternative. The rule of irresponsible finance and the advertisers' press is played out.

These—the credit power and the press—are, we submit, the natural organs of social control in the new world of work. To socialize the agency that decides what wheels shall turn and who shall turn them and the agency that informs the mind of the multitude and evaluates events—is to rectify the business system.

To neglect to do so is to rush through confusion to autocracy and the regime of the machine-gun in the streets.

Your administration, Mr. President, has graven in the thoughts of the people two genetic ideas that taken together can meet the need of our emergency—to wit, the idea of a bank operated primarily not for private but for public ends, and the idea of an organized intelligence qualified to criticize the working plant of a community regarded as a whole.

The regenerative idea that is bedded in our new banking system requires to be deepened so that its principle of public service shall be applied not merely to the currency or exchange problem, but also to the problem of credit. Concerning this matter we hope to address you at another time.

Our immediate request is that you will permit us to develop under your sanction the new social principle that you invoked when you called upon the engineers of the country to survey and register their findings concerning the industrial and commercial equipment of the several states. We venture to hope that you will call upon us to attempt a more intensive and more permanent application of that principle. We desire to establish, first in New York and then in other cities, a political institution, constituted by the free association of competent men in the spirit of the university, to improve the normal operation of the business system.

With faith and loyalty, we beg to sign ourselves

Yours respectfully,

(Signed) H. L. GANTT,
Chairman of the Committee.
Formerly Vice-President of the
American Society of Mechanical Engineers.[1]

---

[1] The other signers of this letter, with their connections, are given in the subscribed list:

CHARLES R. MANN,
President of the American Federation of Teachers of Natural Sciences. Chief of a Bureau created by the Carnegie Foundation at the instance of the United Engineering Societies of the U. S. to suggest improvements in the educational system of the country from the standpoint of engineers.

THE NEW MACHINE

POSTSCRIPT

February 18, 1917.

On rereading this letter the day after its signing, the Executive Committee is prompted to add a word in the hope of

R. B. WOLF,
    Manager Print-paper Mills at Sault Ste. Marie, Canada.
RICHARD A. FEISS,
    Clothing Manufacturer at Cleveland, Ohio.
H. V. R. SCHEEL,
    Manager, Brighton Cotton Mills, Passaic, N. J.
CHARLES FERGUSON,
    Formerly Special Agent of the U. S. Dept. of Commerce in London, Paris and Berlin.
WALTER RAUTENSTRAUCH,
    Secretary of the Committee. Professor of Mechanical Engineering at Columbia University.

The undersigned—being present at the meeting at which it was considered—subscribe our names to the principles embodied in the foregoing letter to the President:

PERCY S. GRANT,
    Rector, Church of the Ascension, N. Y.
THOMAS C. DESMOND,
    Consulting Engineer.
WALTER N. POLAKOV,
    Consulting Engineer.
E. A. LUCEY,
    Factory Manager, South Manchester, Conn.
WM. EUGENE PULIS,
    Factory Manager, Newark, N. J.
LINCOLN COLCORD,
    Journalist.
H. B. BROUGHAM,
    Editorial staff, *Philadelphia Public Ledger.*
JOHN FRANKLIN CROUELL,
    Executive Officer of N. Y. Chamber of Commerce.
FRANK CHAPIN BRAY,
    Editor, *World's Court Magazine.*
HARVEY DEE BROWN,
    Assistant Minister, Church of the Messiah.
FRED R. LOW,
    Editor, *Power.*
FRED E. ROGERS,
    Editor, *Machinery.*

divesting the subject of every complication and reducing it to its essential terms.

What we mean is that the strength of the country, for peace or war, is involved in the operation of its business system. We put our technical knowledge and experience behind the proposition that the business system as it stands is ineffectual simply because it fails to stimulate the creative forces. We have enough technology, but not enough liberty. We lack heart, sense, feeling.

Everybody knows this, and multitudes of discordant efforts are being made to remedy our weakness. But there is no effort that goes deep enough—deep as religion, deep as dying for one's country. The reforms are self-cancelling and futile—as ribs without a spine.

There is need of an all-correlating moral adventure—need of a bright and conspicuous standard to rally the wills of men.

We are thinking not of mental states, but of matters that are entirely objective. Our minds are bent, as your mind is bent, upon the tense pressure of Europe, the rising cost of the necessaries of life, the possibilities of swift mobilization for war, the question of railroads that can run together and of banks that can build cities. We are thinking of nothing but the moment and the most salient facts, and we declare that the problem is too complex for mere circumspection. The disorder is too serious to be dealt with in detail. The disease is too vital for the medication of symptoms.

We should lay the axe to the root of the rotted tree; we should find new bottles for new wine.

One cannot mend a democracy with patches of despotism. The national life is discomposed; if we undertake to meet an emergency with emergency measures only, we shall be still more discomposed. We should have the mental energy and serenity to institute a new power of creation, in the very front of war and waste.

Such are the implications of our request when we ask you to lend us the credit of the Presidency for the establishment of an office in New York and in other cities—to organize the personal forces of those that want to do business in a better way.

We have quite definite plans for the operation of such offices —the enrollment of men for better placement of their abilities, the technical survey of cities, the development of a scientific news-service, of public-service banks, of commercial corporations to lower the cost of food, and so on—but all these plans may be cast aside for other plans.

There is no particular plan that can contain and characterize our purpose. We are men of good will—having each some special faculty that can be applied to the solution of the problem: How shall the nation be made strong? We think it noteworthy that we agree without formulas—and that we are not in doubt what to do.

We offer you our personal service and devotion for the institution of a new and more practical kind of politics.

Signed by

THE EXECUTIVE COMMITTEE OF
"THE NEW MACHINE"

Chapter XX

THE PARTING OF THE WAYS

During 1918 and 1919 Gantt began to fear the beginning of an "economic catastrophe." He thought there was yet time to revise American business and industrial methods and to stave off that disaster if action were taken quickly. His apprehension was due in part to the increasing profits that business was taking beyond what he deemed a just reward for the services rendered. Perhaps no other man of his day saw the inevitable end of the current of affairs so plainly as he. But he did more than foretell the coming event. He told why it was coming and offered a way of escape.

He foresaw that when the impending business panic did come it would be attributed fallaciously to "over-production," and that lessening of production and output would seem to be the course to be taken by manufacturers and producers.

> Over-production has been the bugbear of American business. Our periodic panics have all been laid to this. From time to time we have produced so many goods that it was thought there was no market for them, and the industries have had to shut down. This brought unemployment and poverty, with consequent inability to buy the things we had produced. The workers then had to go ragged because they had produced so many clothes. They had to go barefooted because they had produced so many shoes. They had built so many houses that they

THE PARTING OF THE WAYS

had to live outdoors. Can anyone find an excuse for continuing such a system of industry?

How to curtail production and avoid glutting the market has often been a problem of our business interests. Curtailing productions means shutting down the plant, wholly or in part. The "captain of industry" by this measurement thus becomes too often a captain of idleness. The way to get rich, he discovered, was to quit producing wealth.

We had never had real over-production yet. We have never produced more things than we wanted. All that we have done is to produce more than we could buy. With distribution simplified, that bugbear would be removed. If the time ever comes that we have produced all the things we need, most of us won't mind knocking off work awhile.

His challenging misgivings as to the immediate future he wrote, early in 1919, into a manuscript entitled, "The Parting of the Ways." Several versions of this paper are in existence, showing the care with which he rewrote and revised his language. This paper was published as the first chapter of his last book, *Organizing for Work*. That version is reprinted here with the permission of the publisher, Harcourt Brace & Company. It is at one and the same time a vision and an exhortation.

Modern civilization is dependent for its existence absolutely upon the proper functioning of the industrial and business system. If the industrial and business system fails to function properly in any important particular, such, for instance, as transportation, or the mining of coal, the large cities will in a short time run short of food, and industry throughout the country will be brought to a standstill for lack of power.

It is thus clearly seen that the maintenance of our modern civilization is dependent absolutely upon the service it gets from the industrial and business system.

This system as developed throughout the world had its origin in the service it could and did render the community in which it originated. With the rise of a better technology it was found that larger industrial aggregations could render better and more effective service than the original smaller ones, hence the smaller ones gradually disappeared leaving the field to those that could give the better service.

Such was the normal and natural growth of business and industry which obtained its profits because of its superior service. Toward the latter part of the nineteenth century it was discovered that a relatively small number of factories, or industrial units, had replaced the numerous mechanics with their little shops, such as the village shoemaker and the village wheelwright, who made shoes and wagons for the community, and that the community at large was dependent upon the relatively smaller number of larger establishments in each industry.

Under these conditions it was but natural that a new class of business man should arise who realized that if all the plants in any industry were combined under one control, the community would have to accept such service as it was willing to offer, and pay the price which it demanded. In other words, it was clearly realized that if such combinations could be made to cover a large enough field, they would no longer need to serve the community but could force the community to do their bidding. The Sherman Anti-Trust Law was the first attempt to curb this tendency. It was, however, successful only to a very limited extent, for the idea that the profits of a business were justified only on account of the service it rendered was rapidly giving way to one in which profits took the first place and service the second. This idea has grown so rapidly and has become so firmly imbedded in the mind of the business man of today, that it is inconceivable to many leaders of big business that it is possible to operate a business system on the lines along which our present system grew up; namely, that its first aim should be to render service.

It is this conflict of ideals which is the source of the confusion into which the world now seems to be driving headlong.

THE PARTING OF THE WAYS

The community needs service first, regardless of who gets the profits, because its life depends upon the service it gets. The business man says profits are more important to him than the service he renders; that the wheels of business shall not turn, whether the community needs the service or not, unless he can have his measure of profit. *He has forgotten that his business system had its foundation in service, and as far as the community is concerned has no reason for existence except the service it can render.* A clash between these two ideals will ultimately bring a deadlock between the business system and the community. The "laissez faire" process in which we all seem to have so much faith, does not promise any other result, for there is no doubt that industrial and social unrest is distinctly on the increase throughout the country.

I say, therefore, we have come to the *Parting of the Ways*, for we must not drift on indefinitely toward an economic catastrophe such as Europe exhibits to us. We probably have abundant time to revise our methods and stave off such a catastrophe if those in control of industry will recognize the seriousness of the situation and promptly present a positive program which definitely recognizes the responsibility of the industrial and business system to render such service as the community needs. The extreme radicals have always had a clear vision of the desirability of accomplishing this end, but they have always fallen short in the production of a mechanism that would enable them to materialize their vision.

American workmen will prefer to follow a definite mechanism, which they comprehend, rather than to take the chance of accomplishing the same end by the methods advocated by extremists. In Russia and throughout eastern Europe, the community through the Soviet form of government is attempting to take over the business system in its effort to secure the service it needs. Their methods seem to us crude, and to violate our ideas of justice; but in Russia they replaced a business system which was rotten beyond anything we can imagine. It would not require a very perfect system to be better than what they had, for the dealings of our manufacturers with the Russian business agents during the war indicated that graft was almost the controlling factor in all deals. The Soviet government is not necessarily Bolshevistic nor Socialistic, nor is it political in the ordinary sense, but industrial. It is the first attempt to found a government on industrialism. Whether or not it will be ultimately successful remains to be seen. While the movement is going through its initial stages, however, it is unquestionably working great hardships, which are enormously aggravated by the fact that it has fallen under the control of the extreme radicals. Would it not be better for our business men to return to the ideals upon which their system was founded and upon which it grew to such strength; namely, that reward should be dependent solely upon the service rendered, rather than to risk any such attempt on the part of the workmen in this country, even if we could keep it clear of extreme radicals, which is not likely? *We all realize that any reward or profit that business arbitrarily takes, over and above that to which it is justly entitled for service rendered, is just as much the exercise of autocratic power and a menace to the industrial peace of the world, as the autocratic military power of the Kaiser was a menace to international peace. This applies to Bolshevists as well as to Bankers.*

I am not suggesting anything new, when I say reward must be based on service rendered, but am simply proposing that we go back to the first principles, which still exist in many rural communities where the newer idea of big business has not yet penetrated. Unquestionably many leading business men recognize this general principle and successfully operate their business accordingly. Many others would like to go back to it, if they saw how such a move could be accomplished.

Under stress of war, when it was clearly seen that a business and industrial system run primarily for profits could not produce the war gear needed, we promptly adopted a method of finance which was new to us. The Federal Government took over the financing of such corporations as were needed to furnish the

THE PARTING OF THE WAYS

munitions of war. The financing power did not expect any profit from these organizations, but attempted to run them in such a manner as to deliver the greatest possible amount of goods.

The best known of these is the Emergency Fleet Corporation. It is not surprising that such a large corporation developed in such great haste should have been inefficient in its operating methods, but there are reasons to believe that it will, in the long run, prove to have handled its business better than similar undertakings that were handled directly through the Washington bureaus. It gave us a concrete example of how to build a Public Service corporation, the fundamental fact concerning which is that it must be *financed by public money*. That it has not been more successful is due, not to the methods of its financing, but to the method of its operation. The sole object of the Fleet Corporation was to produce ships, but there has never been among the higher officers of the Corporation a single person, who, during the past twenty years, has made a record in production. They have all without exception been men of the "business" type of mind who have made their success through financiering, buying, selling, etc. If the higher officers of the Fleet Corporation had been men who understood modern production methods, and had in the past been successful in getting results through their use, it is probable that the Corporation would have been highly successful, and would have given us a good example of how to build an effective Public Service corporation.

Mr. William B. Colver, Chairman of the Federal Trade Commission, in the summer of 1917, explained how we might have a Public Service corporation for the distribution of coal. In such a corporation as Mr. Colver outlined, there would be good pay for all who rendered good service, but "no profit." Of course, all those who are now making profits over and above the proper reward for service rendered in the distribution of coal, opposed Mr. Colver's plan, which was that a corporation, financed by the Federal Government, should buy at the mouth of each mine such coal as it needed, at a fair price based on the cost of operating that mine; that this corporation should distribute to the community the coal at an average price, including the cost of distribution. We see no reason why such a corporation should not have solved the coal problem, and furnished us with an example of how to solve other similar problems. We need such information badly, for we are rapidly coming to a point where we realize that *disagreements between employer and employee as to how the profits shall be shared can no longer be allowed to work hardship to the community.*

The chaotic condition into which Europe is rapidly drifting by the failure of the present industrial and financial system, emphasizes the fact that in a civilization like ours the problems of peace may be quite as serious as the problems of war, and the emergencies created by them therefore justify the same kind of action on the part of the government as was justified by war.

Before proper action can be taken in this matter it must be clearly recognized that today economic conditions have far more power for good or for evil than political theories. This is becoming so evident in Europe that it is impossible to fail much longer to recognize it here. The revolutions which have occurred in Europe and the agitation which seems about to create other revolutions, are far more economic than political, and hence can be offset only by economic methods.

The Labor Unions of Great Britain, and the Soviet System of Russia, both aim, by different methods, to render service to the community, but whether they will do it effectively or not is uncertain, for they are revolutionary, and a revolution is a dangerous experiment, the result of which cannot be foreseen. The desired result can be obtained *without a revolution* and by methods with which we are already familiar, if we will only establish real public service corporations to handle problems which are of most importance to the community, and realize that capital like labor is entitled only to the reward it earns.

Inasmuch as the profits in any corporation go to those who finance that corporation, the only guarantee that a corporation

THE PARTING OF THE WAYS 285

is a real public service corporation is that it is financed by public money. If it is so financed all the profits go to the community, and if service is more important than profits, it is always possible to get a maximum service by eliminating profits.

This is the basis of the Emergency Fleet Corporation, and numerous other war corporations, which rendered such public service as it was impossible to get from any private corporations. Realizing that on the return of peace many private corporations feel that they have no longer such social responsibilities as they cheerfully accepted during the war, it should seem that real public service corporations would be of the greatest possible advantage in the industrial and business reorganization that is before us.

We have in this country a little time to think, because economic conditions here are not as acute as they are in Europe, and because of the greater prosperity of our country. But we must recognize the fact that our great complicated system of modern civilization, whose very life depends upon the proper functioning of the business and industrial system, cannot be supported very much longer unless the business and industrial system devotes its energies as a primary object to rendering the service necessary to support it. We have no hesitation in saying that the workmen cannot continue to get high wages unless they do a big day's work. *Is it not an equally self-evident fact that the business man cannot continue to get big rewards unless he renders a corresponding amount of service?* Apparently the similarity of these two propositions has not clearly dawned upon the man with the financial type of mind, for the reason, perhaps, that he has never compared them.

Such a change would produce hardships only for those who are getting the rewards they are not earning. It would greatly benefit those who are actually doing the work.

In order that we may get a clear conception of what such a condition would mean, let us imagine two nations as nearly identical as we can picture them, one of which had a business system which was based upon and supported by the service it rendered to the community. Let us imagine that the other nation, having the same degree of civilization, had a business system run primarily to give profits to those who controlled that system, which rendered service when such service increased its profits, but failed to render service when such service did not make for profits. To make the comparison more exact, let us further imagine a large portion of the most capable men of the latter community engaged continually in a pull and haul, one against the other, to secure the largest possible profits. Then let us ask ourselves in what relative state of economic development these two nations would find themselves at the end of ten years? It is not necessary to answer this question.

I say again, then, we have come to the *Parting of the Ways,* for a nation whose business system is based on service will in a short time show such advancement over one whose business system is operated primarily with the object of securing the greatest possible profits for the investing class, that the latter nation will not be long in the running.

America holds a unique place in the world and by its traditions is the logical nation to continue to develop its business system on the line of service. What is happening in Europe should hasten our decision to take this step, for the business system of this country is identical with the business system of Europe, which, if we are to believe the reports, is so endangered by the crude efforts of the Soviet to make business serve the community.

The lesson is this: *The business system must accept its social responsibility and devote itself primarily to service, or the community will ultimately make the attempt to take it over in order to operate it in its own interest.*

The spectacle of the attempt to accomplish this result in eastern Europe is certainly not so attractive as to make us desire to try the same experiment here. Hence, we should act, and act quickly, on the former proposition.

His final warning on the impending break-up of business and industry was written into a newspaper article in the

THE PARTING OF THE WAYS 287

Sunday *World* of October 12, 1919, and so published only a little longer than a month before his death. Gantt's prediction must now be placed alongside the known events. Business activity began to lessen in February, 1920, reaching a low point in September, 1921. Expansion then set in, continuing on to the boom years of 1927 and 1928. In 1929 business activity began to recede into one of the worst panics the American people have ever experienced. A state of prosperity has not been restored as these lines are written in mid-summer 1934.

Facing these disquieting events, listen to Gantt's prophetic voice, "The system cannot last."

> The business system as at present organized cannot maintain itself much longer. It is entirely out of gear, and it is only revealing more clearly its own condition by trying to run everything and everybody, including the Government. The workers can no longer be forced to work under an arbitrary and autocratic rule. The workers have learned to read statistics and financial reports, and they clamor for a larger share of the tremendous profits made by those who sell. The consumer will not endure forever the mad race of the sellers of things to drive prices up. The system cannot last.

Chapter XXI

THE MOTIVE OF SERVICE

GANTT believed that the democratic operation of business and industry could only be secured when the actuating motives of employers and employees were right. He declared that the motive that both should adopt is the doctrine of service that has been preached in the churches for generations.

His vision, the objective toward which he strove during the latter years of his life, was *universal peace* for all business and industry. He sought to reconcile labor and capital, to reduce friction in employer-employee relations, to remove inequalities in industrial operation, to emphasize useful labor, and to fix rewards—workers' wages, and company earnings—according to essential service rendered.

He wrote of these convictions in regard to service in the May, 1919, issue of *Industrial Management*. He closed with the statement that "*Economic Democracy thus reveals itself as applied Christianity.*" The following excerpts comprise the major part of that paper.

> For over a thousand years the history of the world has been made by two great forces—the church and the state—the church basing its efforts on idealism and moral forces, the state depending almost entirely upon military power. At times these two forces have seemed for a while to cooperate, and then to become antagonistic. Today [1919] they are absolutely distinct, working

THE MOTIVE OF SERVICE

in different fields, with but little ground in common, and a rival claims the middle of the stage, for during the last century there has come into the world another force, which has concerned itself but little with our religious activities, and interested itself in our political activities only insofar as it could make the political forces serve its ends. I speak of the modern business system, based on the tremendously increased productive capacity of the race due to the advance of the arts and sciences. The rapid expansion of this new power has thrown all of our economic mechanism out of gear, and because it has failed to maintain a social purpose, which is common to both of the other forces, produces cross-currents and antagonisms in the community which are extremely detrimental to society as a whole.

One hundred years ago, each family—certainly each community—produced nearly everything needed for the simple life then led.

The village blacksmith and the local mill served the community, which existed substantially as a self-contained unit.

With the growth of the transportation system and grand scale production many of the functions of the local artisans were taken over by the factory, just as the flour mills of Minneapolis supplanted the local mills, which went out of existence.

In the same manner other large centralized industries by superior service drove out of existence small local industries. By reason of improved machinery and a better technology the centralized industries were able to render this superior service, at the same time securing large profits for themselves. Unfortunately for the country at large, those who later came into control of these industries did not see that the logical basis of their profits was service. When, therefore, the community as a whole had come to depend upon them exclusively, they realized their opportunity for larger profits still, and so changed their methods as to give profits first place, oftentimes ignoring the subject of service almost entirely. It is this change of object in the business and industrial system, which took place about the close of the nineteenth century, that is the source of much of the woe that has recently come upon the world. Unless the industrial and business system can rapidly recover a sense of service and grant it the first place, it is hard to see what the next few years may bring forth.

The great war through which we have just passed has done away with political autocracy, apparently forever, but it has done nothing whatever in this country to modify the autocratic methods of the business system, which is a law unto itself and which now accepts no definite social responsibility. This force is controlled by and operated in the interest of ownership, with, in many cases, but little consideration for the interests of those upon whose labor it depends, or for that of the community. We should not be surprised, therefore, that the workman who is most directly affected by this policy is demanding a larger part in the control of industry, especially as the war has taught him, in common with most of us, that the method of operating an industry is more important to the community than the particular ownership of that industry. The result of this knowledge is that the workers throughout the world are striving everywhere to seize the reins of power. Unfortunately, for the world at large, these workers as a rule have no clearer conception of the social responsibility than those already in control. Moreover, having had no experience in operating grand scale industry and business, it is more than likely that their attempt to do so will result disastrously to the community. The industrial system as a whole is thus threatened with a change of control which we can scarcely contemplate with equanimity. We naturally ask if there is any possible relief from the confusion with which we are threatened. We think there is, but not by any of the methods generally advocated "by intellectuals" who are not closely in touch with the moving forces.

One class believes that the answer comes in government ownership and government control of industries. The experience of the world so far does not, however, give much encouragement along these lines, for in some quarters where public

THE MOTIVE OF SERVICE

utilities have to a large extent been run by the government, it is frankly admitted that the government is being run by the business system, which leaves us just where we were, *unless we can get a social purpose into that system,* in which case the need for government ownership would disappear. Is such a thing possible? Unless it can be shown that a business system which has a social purpose is distinctly more beneficial to those who control than one which has not a social purpose, I frankly confess that there does not seem to be any permanent answer in sight. On the other hand, if it can be shown conclusively that a business system operated by democratic methods (and the test of such a system is that it acts without coercion and offers each man the full reward of his labor) is more beneficial to those who lead than the present autocratic system, we have a basis on which to build a modern economic state, and one which we can establish without a revolution, or even a serious jar to our present industrial and business system. In fact, so far as I have been able to put into operation the methods I am advocating, we have very materially reduced the friction and inequalities of the present methods, much to the benefit of both employer and employee.

We are now attempting to make clear that those *who know what to do and how to do it* can most profitably be employed in teaching and training others. In other words, that they can earn their greatest reward by rendering service to their fellows as well as to their employers. It has only been recently that we have been able to get owners and managers interested in this policy, for all the cost systems of the past have recorded such teachers as non-producers and hence an expense that should not be allowed. Now, however, with a proper cost-keeping system supplemented by a man record chart system, we see that they are really our most effective producers.

We have yet to find a single place where these methods are not applicable, and where they have not produced better results than the old autocratic system. Moreover, they produce harmony between employer and employee and are welcomed by both.

In other words, we *have proved in many places that the doctrine of service* which has been preached in the churches as religion is not only good economics and eminently practical, but because of the increased production of goods obtained by it, promises to lead us safely through the maze of confusion into which we seem to be headed, and to give us that industrial democracy which alone can afford a basis for industrial peace.

This doctrine has been preached in the churches for nearly two thousand years, and for a while it seemed as if the Catholic Church of the Middle Ages would make it the controlling factor in the world; but the breaking up of the Church of the Middle Ages into sects, and the advance of that intellectualism which placed more importance upon words and dogma than upon deeds, gave a set-back to the idea which has lasted for centuries. Now, when a great catastrophe has made us aware of the futility of such methods, we are beginning to realize that the present business system needs only the simple methods of the Salvation Army to restore it to health. It is absolutely sound at bottom.

It should be perfectly evident that with the increasing complexity of the modern business system (on which modern civilization depends) successful operation can be attained only by following the lead of those who understand practically the controlling forces, and are willing to recognize their social responsibility in operating them.

Any attempt to operate it by people who do not understand the driving forces is sure to reduce its effectiveness, and any attempt to operate it in the interest of a class is not much longer possible.

For instance, under present conditions the attempt to drive the workman to do that which he does not understand results in failure, even if he is willing to be driven, which he no longer is; for he has learned that real democracy is something more than the privilege of expressing an opinion. We are thus forced into the new economic condition, and, whether we like it or not, will soon realize that only those *who know what to do*

THE MOTIVE OF SERVICE 293

and how to do it will have a sufficient following to make their efforts worth while. In other words, the conditions under which the great industrial and business system must operate to keep our complicated system of modern civilization going successfully, can be directed only by real leaders—men who understand the operation of the moving forces, and whose prime object is to render such service as the community needs.

In 1847, Mr. Lincoln wrote: "To secure to each laborer the whole product of his labor, or as nearly as possible, is a worthy object of any good government. But then the question arises, how can a government best effect this? . . . Upon this subject the habits of our whole species fall into three great classes—useful labor, useless labor and idleness. Of these, the first only is meritorious, and to it all the products of labor rightfully belong; but the two latter, while they exist, are heavy pensions upon the first, robbing it of a large portion of its just rights. The only remedy for this is to, as far as possible, drive useless labor and idleness out of existence."

Attempts are always being made to eliminate the idleness of workmen and useless labor by the refusal of compensation. Unfortunately, however, there has been no organized attempt as yet to force capital to be useful by refusing compensation to idle capital, or to the useless expenditure of same. Capital which is expended in such a manner as to be non-productive, and capital which is not used, can receive interest only by obtaining the same from capital which was productive, or from the efforts of workmen, in either of which cases it gets a reward which it did not earn, and which necessarily comes from capital or labor which did earn it.

Reward according to service rendered is the only foundation on which our industrial and business system can permanently stand. It is a violation of this principle which has been made the occasion for socialism, communism, and Bolshevism. All we need to defeat these "isms" is to reestablish our industrial and business system firmly on the principles advocated by

Abraham Lincoln, in 1847, and we shall establish an economic democracy that is stronger than any autocracy.

Moreover it conforms absolutely to the teachings of all the churches. *Economic democracy thus reveals itself as applied Christianity.* When this has been achieved, and not until then, shall we be on the road to Universal Peace.

In the early autumn of 1919 President Woodrow Wilson called an Industrial Conference to formulate a plan to deal with the condition of national unrest believed to be a "danger greater than war itself." The editors of *Industrial Management* invited a number of industrial engineers to meet and discuss what the Conference should do. As a result of this gathering each engineer present wrote a letter expressing his views. From them the editors of the magazine prepared a composite statement which was sent to the Conference under date of September 26, 1919. Gantt was one of the most active members of this group and one of the signers of the declaration. The document bears unmistakable evidence of his leadership. It follows in full.

A DECLARATION OF PRINCIPLES

NEW YORK CITY, N. Y.
September 26, 1919

TO THE PRESIDENT'S INDUSTRIAL CONFERENCE
WASHINGTON, D. C.

Gentlemen:

You have been called together by President Wilson because, in his own words, "We are now facing a danger greater than war itself," and you have been charged with the responsibility of formulating a plan to meet the present crisis. We deem it our responsible duty to you and to the people of our country to direct attention to those fundamental principles without which industrial peace cannot be permanently established.

The prevalent unrest in industry results from a system which

THE MOTIVE OF SERVICE

permits the acquisition of wealth for which no adequate service has been rendered and tolerates special privilege with the resulting exploitation of men, women, and children.

Great powers have been used arbitrarily and autocratically to exact unmerited profit or compensation by both capital and labor. This policy of exacting profit rather than rendering service has wasted enormous stores of human and natural resources and has put in places of authority those who seek selfish advantages regardless of the interests of the community.

To remove these evils and to secure the conditions that will promote further economic development, we must:

1. Eliminate all unfair privilege of employer or employee, and make business and industry fulfill their responsibilities to the community.
2. Free all industries producing socially necessary commodities or supplying public service from selfish or incompetent autocratic control.

During the war Great Britain and afterward the United States were forced to adopt these principles to achieve victory. It is now clear that for the well-being of our people and the security of civilization, action must be based on these fundamentals in time of peace as well as in time of war. The violation of these principles is responsible for the present crisis in our country and the world.

We hold that these principles are indisputable, and we declare that unless your findings are based upon them, your report and recommendations will be a counsel of confusion and will intensify rather than allay the threatening consequences of the existing crisis.

We, industrial engineers, who make this declaration, are qualified by our constant contact with the problems of modern industry to speak collectively and with authority concerning industrial competence.

Respectfully submitted,[1]

---

[1] The signers were Harrington Emerson, H. L. Gantt, L. V. Estes, Clinton H. Scovell, Walter N. Polakov, G. Charter Harrison, C. E. Knoeppel, L. W. Wallace, Carle M. Bigelow.

A few weeks after this "Declaration of Principles" was made public, the *New York Sunday World* of October 12, 1919, published an interview with Gantt in which he explained the meaning of the Declaration and gave his views as to how the motive of service could be made effective. In that article he first dealt with the question of economic and social organization based on service. He held that it was relatively unimportant who owned the plants and tools of industry.

> In the first place, I have no intention of disturbing the present distribution of ownership. In fact, I am very little concerned with ownership. I do not care who owns the plants and the tools so long as they are properly used and perform adequate service. Ownership is not nearly so important as the proper use of the tools of production. Any plant under proper management can produce in a few years what it is worth and more. If industry is reorganized on a basis of adequate reward for service rendered and all exploitation is eliminated, there will result an equalization of gains which will lead to greater equality in the enjoyment of the good things of life. The important thing is to eliminate autocratic control from industry, which is the cause of profiteering, industrial unrest and of our economic disorganization in general.
>
> Secondly, every plant and industry should be so organized as to induce every man to do his best by giving him adequate reward for his service. I have devised a system of charts which can be used to show at a glance the efficiency of the management, the record of each employee and the record of the machines. By means of these charts every person employed in a plant may be accorded full information concerning himself and others. He may know at all times what he is doing and how he compares with others. If such comparison is unfavorable, he may learn the causes. Our system is not to fire the less efficient, but to teach and train them to greater efficiency. Each man

THE MOTIVE OF SERVICE 297

or woman is rewarded and promoted to positions of responsibility on the basis of his or her record. This is industrial democracy properly understood, democracy based not on opinions and arbitrary will but on facts and merits.

By placing only those in authority who know what to do and how to do it we may eliminate the arbitrary control of ownership. Industry will then be managed by those who are trained for the job. By rewarding only for services rendered we shall stimulate a lowering of the costs of production and reduce living costs. What I have in mind is not merely a change in the method of rewarding employees, but a complete change in our present system of cost accounting. Under the cost keeping in general use every article is made to bear not only the expense of producing it but also a share of the expenses incurred in maintaining parts of the plant which remain unproductive. Such a system of accounting loads costs with the dead weight of idle machinery and tools and affords an excuse for high prices. This system of cost-keeping makes possible the claim of idle capital for a reward for which it can show no service, and has been devised mainly in response to the demands of ownership. If our system of accounts were changed, the expense of leaving capital in idleness would become so evident that our manufacturers would be impelled to find out the causes and to eliminate them as quickly as possible. In fact, those manufacturers who have once applied this principle, either consciously or unconsciously, have never given it up and have become leaders of industry. Henry Ford can be cited as an illustration.

Gantt outlined in brief how the application of the motive of service would change the system of distribution, and advocated a reorganization of retail trade:

My idea is that production and distribution are two different processes subject to different economic laws. At present, the selling end of every business dominates the production end. Everybody knows that money is made in selling, not in manufacturing. Every energetic, capable young man who has any ambition is eager to enter the sales department in order to make more money. The men who are most successful are those who can sell most at highest prices. Our bankers encourage those business men who can show the greatest selling results. It has now gone so far that many manufacturers care but little to reduce manufacturing cost. They feel they can make their profits by raising prices. When we industrial engineers tell manufacturers how they can reduce costs they frequently meet us with the statement that it is easier for them to raise the price of their articles. In fact, there is little inducement for them to organize their plants in order to reduce costs.

Well, this discrepancy between production and distribution is the chief cause of our high prices. We must separate the two processes. We must put industry in the hands of those who will organize it for production at lowest possible cost. And we must induce them to do so by rewarding them for achieving such results.

Distribution at present is entirely too expensive, and our whole business system is so organized as to encourage this by permitting people to make profits by selling at prices without regard to what the costs of production are. Production is and should be individualistic, but distribution should be organized so as to make selling collective. I am in favor of an organization similar to the Cartel system. If in every industry we had a committee consisting of representatives of producers, distributors and consumers, such a committee could fix prices with due regard to supply and demand. At present when we do not know what the supply of or demand for an article is, the so-called law of supply and demand is a sham. But a committee which had all the facts at its command could fix prices adequately, so as to induce producers to increase production while stimulating them at the same time to decrease costs. Some such plan was suggested two years ago by Mr. Colver of the Federal Trade Commission for distributing the supply of coal, but his plan was not given a trial.

There is no doubt that our system of retailing needs reorgani-

THE MOTIVE OF SERVICE 299

zation if prices are to be reduced. There is no reason why six or seven separate wagons should call to deliver milk to residents of one end of the same block. Perhaps, consumers' co-operative societies would best solve this phase of the problem.

As to the initiative that might effect these reorganizations, Gantt distrusted governmental interference, but had faith in the people and in education. "I have no faith in governmental machinery, as government is at present organized. The only practical plan I can suggest is for the owners of property to realize the truth of such views as I have expressed and to follow the course I outlined. . . . Education of the people to a true understanding of the situation is our most important task.

Appendix

BIBLIOGRAPHY OF WRITINGS OF HENRY LAURENCE GANTT

THE greater number of the items in this bibliography are in publications readily available for library research. Others, manuscripts, pamphlets, and articles in less accessible publications, have a reference to the collection where the material is on file. These are the libraries of the following individuals: L. P. Alford, Colonel W. L. Conrad, F. D. Manning, F. J. Miller, David B. Porter, Charles N. Underwood.

BOOKS:
    Work, Wages, and Profits, 1st edition (1910)
      Engineering Magazine Company.
    Work, Wages, and Profits, 2nd edition (1913)
      Engineering Magazine Company.
    Industrial Leadership (1916)
      Yale University Press.
    Organizing for Work (1919)
      Harcourt, Brace and Howe.

PAPERS AND DISCUSSIONS IN THE "TRANSACTIONS OF THE AMERICAN SOCIETY OF MECHANICAL ENGINEERS":
    "Steel Castings," Vol. XII.
    "Recent Progress in the Manufacture of Steel Castings," Vol. XV.
    "Discussion, 'A Piece Rate System'," Vol. XVI.
    "A Bonus System of Rewarding Labor," Vol. XXIII.

APPENDIX 301

"Discussion, 'Gift Proposition for Paying Workmen'," Vol. XXIV.
"Discussion, 'The Machine Shop Problem'," Vol. XXIV.
"A Graphical Daily Balance in Manufacture," Vol. XXIV.
"Modifying Systems of Management," Vol. XXV.
"Discussion, 'Is Anything the Matter with Piece Work?'," Vol. XXV.
"Discussion, 'On the Art of Cutting Metals'," Vol. XXVIII.
"Discussion, 'College and Apprentice Training'," Vol. XXIX.
"Discussion, 'Industrial Education'," Vol. XXIX.
"Training Workmen in Habits of Industry and Cooperation," Vol. XXX.
"Mechanical Engineering in the Textile Industry," Vol. XXXII.
"Discussion, 'Locomotive Handling at Terminals'," Vol. XXXII.
"Discussion, 'Present State of the Art of Industrial Management'," Vol. XXXIV.
"Discussion, 'Industrial Service Work in Engineering Schools'," Vol. XXXVI.
"Measuring Efficiency," Vol. XXXVI.
"The Relation between Production and Costs," Vol. XXXVII.
"Tribute to Frederick W. Taylor," Vol. XXXVII.
"Productive Capacity a Measure of Value of an Industrial Property," Vol. XXXVIII.
"Discussion, 'Industrial Management'," Vol. XXXVIII.
"Discussion, 'How Does Industrial Valuation Differ from Public Utility Valuation'," Vol. XXXVIII.
"Discussion, 'Graphical Control on the Exception Principle for Executives'," Vol. XXXVIII.
"Discussion, 'Topical Discussion on Management'," Vol. XXXVIII.
"Expenses and Costs," Vol. XXXIX.
"Efficiency and Democracy," Vol. XL.
"Discussion, 'Non-Financial Incentives'," Vol. XL.
"Discussion, 'Industrial Organization as It Affects Executives and Workers'," Vol. XL.

ARTICLES, ADDRESSES, REPORTS, AND MANUSCRIPTS:

"The Efficiency of Fluid in Vapor Engines"
Van Nostrand's Engineering Magazine, 1884.

"Economical Heating with Oil"
Iron Age, June 21, 1894.

"Regenerative Garbage Cremators"
Engineering Record, July 14, 1894.

"Gaseous Fuels"
Cassier's Magazine, November, 1895.

"Synthesis of Carbon Compounds"
American Manufacturer, May 8, 1896.

"A New High Temperature Furnace"
Journal of the Franklin Institute, November 17, 1896.

"Economical Gas Heating"
American Manufacturer, October 30, 1896.

"Casting of Large Ingots"
Manuscript, 1901 (Conrad collection).

"Report to Bethlehem Steel Company on Defects in Ingots Cast in a Sand Lined Mold"
August 21, 1899; July 26, 27, 1900; August 1, 27, 28, 1900; June 4, 1901 (Conrad collection).

"A New System of Rewarding Machine Shop Laborers"
Cassier's Magazine, 1902.

"Machine Shop Practice"
"Nine engineering reports with blueprints and forms on American Locomotive Company," January 15-August 18, 1902 (Underwood collection).

"The Bonus System"
World's Work, March, 1902.

"Belting"
Manuscript, March 25, 1902 (Conrad collection).

"The Bonus System"
Review of Reviews, September, 1902.

"Application of Scientific Methods to the Labor Problems"
American Machinist, Vol. 27, 1904.

"Exact Time Study as the Basis of Pay for Work Done"

APPENDIX 303

Address, National Metal Trades Association Convention, Philadelphia, March, 1904.

"Principles of Management"
Unsigned manuscript, July 5, 1904 (Underwood collection).

"Principles of Management"
Address, International Congress of Arts and Sciences, St. Louis Fair, 1904.

Gantt's original reports, records, and forms at Sayles Bleacheries, 1904 and following.

"The Compensation of Labor"
Engineering Magazine, March, 1905.

"Economical Utilization of Labor"
Engineering Magazine, February, 1907.

"Economical Utilization of Labor"
Machinery, April, 1907.

Gantt's original reports, records, and forms at Stokes & Smith, 1907.

"Task and Bonus in Management"
Stevens Institute Indicator, April, 1908.

"Continuous Bleaching"
Manuscript, 1909 (Conrad collection).

"Compensation of Workmen and Efficiency of Operation"
Engineering Magazine, February, 1910.

"Compensation of Workmen and Efficiency of Operation"
Engineering Magazine, March, 1910.

"Management of a Bleachery"
Manuscript, 1910.

"Compensation of Workmen and Efficiency of Operation"
Engineering Magazine, April, 1910.

"The Engineer as a Manager"
Stevens Institute Indicator, April, 1910.

"Production Increasing Methods—Training Workmen"
Address, National Association of Cotton Manufacturers, Boston, April, 1910.

"Compensation of Workmen and Efficiency of Operation"
Engineering Magazine, May, 1910.

"Compensation of Workmen and Efficiency of Operation"
  Engineering Magazine, June, 1910.
"Belting"
  Manuscript, November 8, 1910 (Conrad collection).
"Compensation of Workmen and Training of Workmen"
  Address, Harvard School of Business Administration, December, 1910.
"The Straight Line to Profit"
  System, February, 1911.
"Practical Application of Scientific Management"
  Engineering Magazine, April, 1911.
"German Thoroughness"
  Manuscript, April, 1911.
"Task Work the Basis of Proper Management"
  Address, Tenth Annual Convention, National Machine Tool Builders, October 10, 1911 (Underwood collection).
"Problems of Industrial Efficiency"
  Industrial Engineering and Engineering Digest, Vol. 9, 1911.
"An Explanation of Scientific Management"
  Article by F. W. Taylor and H. L. Gantt, *Canadian Manufacturer*, 1911.
"The Task and the Day's Work"
  Address, Harvard College Conference on Scientific Management, October, 1911 (Underwood collection).
"The Engineer as a Manager"
  Bulletin of the Society for the Promotion of Engineering Education, June, 1912.
Conference on the "Taylor System"
  Amos Tuck School of Administration and Finance, Dartmouth College, October, 1912.
"Giving the Worker His Share"
  New York Evening Post, December 6, 1912.
Testimony in hearing on the "Taylor and Other Systems of Shop Management," H. Res. 90, 1912.
Gantt's original reports, records, and forms at Cheney Brothers, 1912 and following.

APPENDIX 305

Set of forms used at Lewiston Bleachery and Dye Works, 1913 (Underwood collection).

"Task and Bonus"
  Address, Remington Typewriter Work Offices, October 15, 1913 (Alford collection).

"Duties and Responsibilities of the Coming Era"
  Address as Presiding Officer of the American Society of Mechanical Engineers. Reported in *New York Evening Post,* June 18, 1914.

"Rules of Management"
  Manuscript, August 15, 1914 (Conrad collection).

"Task Management"
  Manuscript, October 19, 1914 (Conrad collection).

"The Industrial Engineer"
  Manuscript, November 30, 1914 (Conrad collection).

"The Value of Non-Productive Labor"
  Industrial Engineering and Engineering Digest, 1914, Vol. 14.

"Efficiency Systems and Labor"
  Final Report and Testimony Submitted to Congress by the Commission on Industrial Relations, Vol. I, 1914.

"Industrial Leadership"
  Manuscript, February 26, 1915 (Conrad collection).

"Principles Underlying a Proper Manufacturing System"
  Manuscript, May 27, 1915 (Conrad collection).

"Organization of a Manufacturing Plant"
  Manuscript, June 17, 1915 (Conrad collection).

"Modern Methods of Training Workmen"
  Manuscript, August 7, 1915 (Conrad collection).

"How to Create Industrial Leaders"
  Engineering Magazine, December, 1915.

"The Effect of Idle Plant on Costs and Profits"
  Annals of American Academy of Political and Social Science, 1915.

"Production and Sales"
  Engineering Magazine, January, 1916.

"Complete Time Study Observation on 'Margin Stop Rack'"
  Manuscript, March 18, 1916 (Alford collection).
"Industrial Leadership"
  Address, Johns Hopkins University, April 14, 1916 (Underwood collection).
"The Importance of Leadership"
  Engineering Magazine, May, 1916.
"Method of Making Tools"
  Manuscript, March, 1916 (Alford collection).
"Engineering Schools and Industrial Methods"
  Engineering Magazine, May, 1916.
"What Is Preparedness?"
  Engineering Magazine, September, 1916.
"The New Machine"
  Three articles reprinted from *Philadelphia Public Ledger*, December 10, 1916 (Underwood collection).
"Report on Screw Machine Data, Remington Typewriter Works"
  Manuscript, January 8, 1917 (Alford collection).
"Belting"
  Manuscript, February 16, 1917 (Conrad collection).
Letter to the President of the United States with reference to The New Machine
  February 17, 1917 (Underwood collection).
Tool Power, Not Wealth, Supplies Sinews of War
  Philadelphia Public Ledger, April 10, 1917 (Alford collection).
"War for Democracy"
  Evening Mail, April 24, 1917 (Alford collection).
"Making Good versus Making Records"
  Engineering Magazine, May, 1917.
"The Basis of Manufacturing Costs"
  Engineering Magazine, June, 1917.
"Efficiency Expert's View of the National Outlook"
  New York Evening Post, June 4, 1917.
"Business and War"
  New York Evening Mail, August 28, 1917.

APPENDIX 307

"Manufacturing Methods"
   Engineering Magazine, September, 1917.
"Business and War"
   Bankers Journal, October, 1917.
"Hard Days, Troubled Waters and a Leaky Fighting-ship"
   Philadelphia Public Ledger, November 2, 1917 (Alford collection).
"Industrial Leadership and Its Relation to Production and Scientific Distribution"
   Journal of the American Bankers Association, November, 1917.
"Tools the Controlling Power of the World"
   Newspaper article, 1917 (Alford collection).
"Private Price-fixing with a Change of Heart"
   Philadelphia Public Ledger, November 3, 1917.
"Report on Shipping Board Problems"
   Manuscript, 1917 (Conrad collection).
"Drafting Industry"
   World Court Magazine, 1917 (Alford collection).
"Danger of Overproduction"
   Baltimore American, February 15, 1918.
"The Manager and His System"
   Manuscript, February 19, 1918 (Underwood collection).
"Handling Defective Material Produced in Manufacturing"
   Manuscript, March 20, 1918 (Conrad collection).
"Building Earning Power"
   Factory, March, 1918.
"War Production and Transportation"
   Bankers Journal, March, 1918.
"Passing the Buck"
   Industrial Management, May, 1918.
Industrial Democracy with Maximum Production as a Means to War-winning and Cost-paying Efficiencies
   Interview with Gantt, *New York Sunday World,* June 30, 1918.
"The New Business"
   New York Tribune, November 18, 1918.

"Keeping the War Won"
   Address, Engineer's Club of Baltimore, December 11, 1918 (Underwood collection).
"Keeping the War Won"
   Manuscript, November 21, 1918 (Conrad collection).
"Efficiency and Idleness"
   Industrial Management, November, 1918.
"Works Management"
   Manuscript, 1918 (F. J. Miller collection).
Testimony before the Federal Trade Commission with Reference to Fixing the Price of Newsprint, 1918.
"American Finance and Production"
   Newspaperdom, January 9, 1919 (Underwood collection).
"Democratic Shop Methods"
   Address, plant of the De Laval Steam Turbine Co., January 29, 1919 (Conrad collection).
"Report on Standard Symbols for Idle Man Charts and Man Record Charts"
   Manuscript, March 31, 1919.
"Democracy in Production"
   Manuscript, May 7, 1919 (Conrad collection).
"Democracy in the Shop"
   Manuscript, May 8, 1919 (Conrad collection).
"The Parting of the Ways"
   Manuscript, May 5, 1919 (Conrad collection).
"Storekeeping"
   Manuscript, Smith Premier Works, January 25, 1919 (Alford collection).
The Religion of Democracy
   Industrial Management, May, 1919 (Underwood collection).
"Organizing for Work"
   Industrial Management, August, 1919 (Underwood collection).
"A Declaration of Principles"
   Open letter to the President's Industrial Conference, September 26, 1919 (Alford collection).

APPENDIX 309

"The Influence of Executives"
  Annals of the American Academy of Political and Social Science, Philadelphia, September, 1919 (Underwood collection).

"The Engineer's Formula for Reconciling Capital and Labor on a Basis of Equity"
  Interview with Gantt, *New York Sunday World*, October 12, 1919.

"Declaration of Principles"
  Montclair Times, November 1, 1919 (Alford collection).

"Democracy in Management"
  Manuscript, November 8, 1919.

"Starting a Job"
  Manuscript, November 12, 1919 (Conrad collection).

Gantt's original records, reports and forms at J. H. Williams and Company, 1919 (F. D. Manning collection).

"Organizing for Production"
  Address prepared for the Industrial Extension Institute, 1919 (Underwood collection).

NOT DATED:

"Efficiency Engineers and Production"
  Manuscript (F. J. Miller collection).

"A Material Handling and Transport Organization"
  Manuscript (Conrad collection).

"Industrial Efficiency"
  Manuscript (Conrad collection).

"Industrial Leadership"
  Southern Commercial Congress.

"Mechanical Engineer and Cost Accountant."

"Outline of Time Study."

Early Gantt Charts (Porter collection).

INDEX

Adams, K. E., 107
Afleck, David, 107
Aldrich, William S., 44
Allan, Colonel William, 19, 23, 26, 29, 131
American Engineering Council, 248
American Federation of Labor, 109
American Locomotive Co., 104, 111
American Society of Mechanical Engineers, 75, 78, 96, 108, 131, 138, 173, 239, 265, 301
American Steel Car Wheel Co., 62
Ancestry, 6
Armor plate patent, 63
Arnold, Solon, 39
Ayers, H. B., 104

Baker, Charles Whiting, 48
Ball-grinding machine, 66
Barth, Carl G., 63, 67, 68, 69
Belgian refugees, 188
Belgian relief, 188
Benjamin Atha and Illingworth Co., 62
Bethlehem Iron Co., 81
Bethlehem Steel Co., 81, 86, 102
Bibliography, 300
Bigelow, Carle M., 117, 295
Birthplace, 5
Boiling machine, 115, 124
Bonus chart, 99, 117
Bonus Club, 127
Bonus plan, 85
Bray, Frank Chapin, 275
Brighton Mills, 111, 143, 144
British Institution of Mechanical Engineers, 75
Brougham, H. B., 275
Brown, Reverend Francis A., 47
Brown, Harvey Dee, 275

Brunker, Albert R., 194, 241
Bryan, Mrs. William J., 38
Buscher, T. W., 107
Butterworth, James F., 186, 188, 240

Chart, cost, 164
   daily balance, 100
   Gantt, 106, 207
   idleness expense, 183
   percentage, 118
   production, 164
   red-and-black, 99, 117
Chase, George H., 63
Cheney Brothers, 166
Cheney, Horace B., 166
Chute, Gantt, 115, 121
Clark, Wallace, 156, 200, 213
Colcord, Lincoln, 275
Colver, William B., 181, 283
Conrad, William L., 81, 107
Copley, Frank B., 128
Cost chart, 164
Cox and Sons Co., 65
Crouell, John Franklin, 275
Crozier, General William, 193, 207

Day, Charles, 96, 110, 189
Democracy, definition of, 252
   in industry, 251, 259
Depression, prediction of, 278
Desmond, Thomas C., 275
Differential piece-work, 91
Distribution, system of, 297
Dornin, George A., 64, 107, 226

Earl, E. P., 87, 94
Egan, Patrick, 72
Emergency Fleet Corp., 193, 198
Emerson, Harrington, 189, 238, 295

312 INDEX

Employers' associations, 257
Ernst, Alfred F., 143
Estes, L. V., 295
Evans, Marshall S., 240

Federal Trade Commission, 177
Feiss, Richard A., 275
Fenner, D. C., 84, 90, 102, 107
Ferguson, Charles, 264, 265, 275
Frankford Arsenal, 189, 209
Franklin, Benjamin A., 71
Freminville, Charles de, 241
Frey, John P., 109

Gantt, Charles, 14
Gantt, Edward, 7
Gantt, Henry Laurence, American Locomotive Co., 104, 111, 208
  American Steel Car Wheel Co., 62
  anecdotes, 11, 12, 15, 36, 37, 43, 48, 50, 54, 71, 73, 110, 126, 135, 136
  appreciation of, 67, 83, 96, 104, 132, 136, 138, 139, 166, 197, 202, 238, 239, 240, 241, 242, 243
  armor-plate invention, 63
  ball-grinding invention, 66
  Belgian relief, 188
  Benjamin Atha and Illingworth Co., 62
  Bethlehem Steel Co., 81, 86, 102
  Brighton Mills, 111, 143, 144, 208
  charts, 99, 100, 106, 117, 118, 164, 183, 207
  Cheney Brothers, 166, 208
  chute, 115, 121
  consulting engineer, 61, 108
  Cox and Sons Co., 65
  death of, 236
  declaration of principles, 294
  devotion to children, 51
  Education, 17, 35
  Emergency Fleet Corp., 193, 198
  experiences with strikes, 127
  Federal Trade Commission testimony, 177
  Frankford Arsenal, 189, 209
  heating furnaces, 65, 66

Gantt, Henry Laurence—(*Continued*)
  high-speed steel, 82
  homes and family, 46
  hunting and fishing, 50, 234
  industrial democracy, 251
  ingot mold, 64
  inventions, 63, 121
  Johns Hopkins University, 35
  Kier, 115, 124
  labor policies, 109, 138, 251, 255, 256
  Liquid Carbonic Co., 194
  love of the beautiful, 54
  management methods, 112, 116, 159, 162, 167
  management results, 88, 118, 127, 136, 143, 144, 160, 161
  McDonogh School, 17, 186, 224
  membership in societies, 75
  memorial medal, 243
  Midvale Steel Works, 62, 71, 79
  military studies, 186
  Montclair home, 49
  motive of service, 288
  Navy report, 189
  Ordnance Bureau, 193
  overhead expense practice, 173
  patents, 63, 121, 124
  piling machine, 115, 121
  Pine Island farm, 224
  plantation home, 5
  Portland Co., 111
  prediction of depression, 278
  preparedness for war, 190
  recreation, 224
  relation with Taylor, 128
  Remington Typewriter Co., 109, 149, 158, 208
  Robins Conveying Belt Co., 111
  Sayles Bleacheries, 107, 126, 208
  scientific management, 78
  Shipping Board, 193
  shop-foreman, 70, 73
  Simonds Rolling Machine Co., 65, 81
  Stevens Institute of Technology, 35
  Stokes & Smith Co., 112
  Tabor Manufacturing Co., 111

INDEX 313

Gantt, Henry Laurence—(*Continued*)
  task and bonus, 85
  The New Machine, 264
  training workmen, 108, 131, 138
  tribute to Taylor, 134
  war activities, 185, 192
  War Industries Board, 193
  washing-machine, 115
  Westinghouse Electric and Manufacturing Co., 127
  Williams and Wilkins, 173
  Williamson Brothers Co., 111
Gantt machine record chart, 217
Gantt man-record chart, 219
Gantt, Margaret Heighe, 10, 48
Gantt, Mary, 11
Gantt, Mary Snow, 46, 53
Gantt, Peggy, 48
Gantt progress chart, 218
Gantt, Thomas, 7
Gantt, Dr. Thomas Compton, 8, 11
Gantt, Virgil, 6, 9, 12, 15
Gardner, R. S., 127
Gates, T. P., 107
General Electric Co., 72
Gilbreth, Frank, 239
Gilbreth, Lillian, 48, 239
Gilman, D. C., 14, 39
Glass furnaces, 65
Glenlyon Dye Works, 117
Going, Charles B., 43
Grammer, Reverend Carl E., 37, 54
Grant, Percy S., 275
Graphical daily balance, 100

Harrison, G. Charter, 295
Heating furnaces, 65, 66
Higgins, M. P., 96
High-speed steel, 82
Humphreys, Alex C., 139
Hunting and fishing, 50, 234

Idleness, 258
Idleness expense, 176
Idleness expense chart, 183
Ingot mold, 64
Inventions, 63, 121, 209

Jackson, Steuart, 230
Jackson, General Thomas J. ("Stonewall"), 19, 31, 185
Jelleme, W. O., 144
Johns Hopkins University, 35
Johnson, John, Jr., 22, 24

Kent, William, 138
Kier, 115, 124
Kimball, Dexter S., 86
Knoeppel, C. E., 295
Knox, S. L. Griswold, 68

Labor policies, 108, 109, 131, 138, 251, 255, 256
Labor unions, 255
Lauer, Conrad N., 110, 139
Lee, General Robert E., 29
Lehigh University, 68
Libby, Samuel H., 71
Linderman, Robert P., 81
Liquid Carbonic Co., 194
Low, Fred R., 275
Lucey, E. A., 107, 275
Lyle, D. C., 238
Lytle, Charles W., 91

Machine record chart, 217
Machine, the new, 264
Management methods, 112, 116, 159, 162, 167
Management results, 88, 118, 127, 136, 143, 144, 160, 161
Mann, Charles R., 274
Manning, F. D., 157, 195, 211
Man-record chart, 219
Maryland plantation, 5, 131
Massachusetts Institute of Technology, 113
Maury, D. H., 42
McDonogh, John, 20, 27
McDonogh School, 7, 186, 224
Medal, Gantt memorial, 243
Midvale Steel Works, 62, 71, 79
Miller, Fred J., 131, 132, 133, 149, 151, 163, 240
Mitchell, Harvey F., 43
Money, 253

INDEX

Montclair home, 49
Montgomery, Colonel George, 189
Morris, Charles D., 37
Morton, Henry, 40
Motive of service, 288
Mullany, John M., 107

Oakleigh, 49
Ordnance Bureau, 193
Ownership, 296

Passano, Edward B., 173, 201
Pelot, Colonel J. H., 190
Percentage chart, 118
Pettigrew, Thomas, 38
Piez, Charles, 138
Piling machine, 115, 121
Pine Island farm, 224
Polakov, Walter N., 266, 275, 295
Poole and Hunt, 61
Porter, David B., 209, 221
Portland Co., 111
Preparedness for War, 190
Preston, Margaret J., 27, 31
Principles, Declaration of, 294
Production chart, 164
Progress chart, 218
Pulis, W. E., 107, 242, 275

Rautenstrauch, Walter, 275
Read, C. O., 113
Red-and-black chart, 99, 117
Reno, H. P., 107, 129, 136
Reid, Harry Fielding, 38
Remington Typewriter Co., 109, 149, 158
Robins Conveying Belt Co., 111
Roe, Joseph W., 167, 207
Rogers, Fred E., 275

Safety and production report, 249
Sayles Bleacheries, 107, 126, 224
Sayles, F. A., 113
Sayles, F. C., 113
Sayles, W. F., 112
Scheel, H. V. R., 139, 275
Schneider, Herman, 201
Scientific management, 78

Scovell, Clinton H., 295
Selling prices, 262
Service, motive of, 288
Shaw, R. D., 107
Sheffield Scientific School, 73, 134
Shipping Board, 193
Simonds Rolling Machine Co., 65, 81
Sinclair, George M., 67, 78
Slide rule, 67
Snow, Mary Eliza, 46
Snyder, R. J., 87
Society of Naval Architects and Marine Engineers, 76
Steuart, Mary Jane, 9
Stevens Institute of Technology, 35
Stokes & Smith Co., 112
"Stonewall" Jackson, 19, 31, 185
Supervisory expenses, 262
Suplee, Henry Harrison, 195

Taber, Margaret Gantt, 243
Tabor Manufacturing Co., 111
Tasks and bonus, 85
Taylor, Frederick W., 63, 66, 67, 68, 69, 79, 80, 81, 84, 91, 108, 112, 128, 132, 134
Taylor-White tool steel, 82
Thompson, General John T., 197
Thurston, Robert H., 42
Training workmen, 108, 131, 138
Tucker, John Randolph, 30
Twelve-hour shift report, 249
Two-bin plan, 112

Underwood, C. N., 107
U. S. Navy report, 189

Veblen, Thorstein, 264
Verein Deutscher Ingenieure, 75
Volkhardt, C. E., 107

Wages, 262
Wallace, L. W., 295
War Industries Board, 193
Washington College, 29
Waste in industry report, 249
Wealth, 254
Wentworth, R. A., 107, 110, 242

INDEX

Westinghouse Electric and Manufacturing Co., 127
White, Maunsel, 82
Williams and Wilkins Co., 173
Williamson Bros. Co., 111
Wilson, Woodrow, 269, 294

Wolf, Robert, 93, 275
Wood, K. F., 112
Workmen training, 108, 131, 138
World War, 142, 185

Zoltaszck, J., 222